AF358627

Magnetic Material Modelling of Electrical Machines

Magnetic Material Modelling of Electrical Machines

Editors

Anouar Belahcen
Armando Pires
Victor Fernão Pires

MDPI • Basel • Beijing • Wuhan • Barcelona • Belgrade • Manchester • Tokyo • Cluj • Tianjin

Editors
Anouar Belahcen
Aalto University
Finland

Armando Pires
Polytechnic Institute of
Setúbal
Portugal

Victor Fernão Pires
Instituto Politécnico de
Setúbal
Portugal

Editorial Office
MDPI
St. Alban-Anlage 66
4052 Basel, Switzerland

This is a reprint of articles from the Special Issue published online in the open access journal *Energies* (ISSN 1996-1073) (available at: https://www.mdpi.com/journal/energies/special_issues/Magnetic_Material_Modelling_Electrical_Machines).

For citation purposes, cite each article independently as indicated on the article page online and as indicated below:

LastName, A.A.; LastName, B.B.; LastName, C.C. Article Title. *Journal Name* **Year**, *Volume Number*, Page Range.

ISBN 978-3-0365-6354-1 (Hbk)
ISBN 978-3-0365-6355-8 (PDF)

Contents

Editorial

Magnetic Material Modelling of Electrical Machines

Anouar Belahcen [1,*], Armando Pires [2,3] and Vitor Fernão Pires [2,4]

[1] Department of Electrical Engineering and Automation, Aalto University, P.O. Box 15500, 00076 Espoo, Finland
[2] SustainRD, EST Setubal, Polytechnic Institute of Setúbal, 2914-508 Setúbal, Portugal
[3] CTS-UNINOVA, 2829-516 Caparica, Portugal
[4] INESC-ID, 1000-029 Lisboa, Portugal
* Correspondence: anouar.belahcen@aalto.fi; Tel.: +358-504602366

Citation: Belahcen, A.; Pires, A.; Pires, V.F. Magnetic Material Modelling of Electrical Machines. *Energies* **2023**, *16*, 654. https://doi.org/10.3390/en16020654

Received: 20 December 2022
Accepted: 21 December 2022
Published: 5 January 2023

1. Introduction

As Guest Editors of this Special Issue, it was our responsibility to ensure that the contributions to the issue related to the extensive field of electromechanical energy conversion, with a special focus on the design, materials, and modeling of electrical machines. This way, electromechanical energy conversion was addressed in the context of electrical motors, generators, and actuators. One fundamental aspect related to this conversion process is that its efficiency and effectiveness depend on device design and the materials used. In this context, several aspects should be considered. One is the design process of the referred devices, which is usually carried out through extensive numerical field computations. Nevertheless, to ensure the accuracy of these computations, the quality of the material models that are adopted must be taken into consideration. Another aspect that requires attention is the modeling of properties such as hysteresis, alternating and rotating losses, and demagnetization. The characterization of the materials and their dependency on mechanical qualities such as stresses and temperature must also be considered. Finally, the design of the drives associated with electrical machines, when required, is another aspect that needs to be considered for the development of the optimal global system. The Special Issue addresses these aspects, contributing to a greater optimization of these kinds of systems.

A brief description of the published papers in this Special Issue is presented in the next section as a concise overview of their content.

2. A Short Review of the Contributions in This Issue

A comparison of seven different methods for determining effective magnetization curves is presented by Tomasz Garbiec and Mariusz Jagiela [1] concerning the use of a field-circuit multi-harmonic model of an induction machine. The accuracy of each method was evaluated by calculating the performance characteristics of a solid-rotor induction machine. The analyses showed that the best practical approach, even for the multi-harmonic case, is to express the effective magnetic permeability as the ratio of the amplitudes of the fundamental harmonics of the magnetic flux density and the magnetic field strength, assuming a sinusoidal variation in the latter.

A new surrogate optimization routine for the design of a direct online (DOL) squirrel cage induction motor is proposed by Aswin Balasubramanian et al. [2]. The motor geometry is optimized to maximize its electromagnetic efficiency while respecting its constraints, such as the output power and the power factor. This novel, efficient, and reliable surrogate optimization routine can be applied to multiple design problems. The proposed clustering technique used in the routine allows for surrogate model accuracy while exploring promising subsets of the design variables range.

Jordi Garcia-Amorós et al. proposed [3] a novel two-phase linear hybrid reluctance actuator with a double-sided segmented stator using laminated U cores and an internal mover with permanent magnets. The permanent magnets are arranged to increase the thrust force of a double-sided linear switched reluctance actuator of the same size, this being the main objective together with achieving a low detente force. A comparative study between the proposed linear hybrid reluctance actuator and a linear permanent magnet actuator of the same size is also presented.

A novel 6/17 E-shaped stator tooth flux switching permanent magnet machine with an added magnet in the upper apex of each dummy slot was proposed by J. Cao et al. [4]. This proposal was derived from the conventional 6/17 E-shaped stator tooth FSPM machine. This structure exhibits several advantages, such as lower torque ripple and lower THD of phase back-EMF.

The importance of electrical machines and their selection in the context of renewable energy applications associated with important infrastructures, such as the case of a water pump, was addressed by Pires et al. [5]. Thus, an 8/6 SRM machine associated with a multilevel converter with fault-tolerant capability was proposed. This converter was designed with the purpose of reducing the cost of these systems and, at the same time, improving their reliability.

Another aspect of the machine addressed in this Special Issue was the problem of torque pulsations. In this ambit, A. Anuchin et al. proposed [6] a continuous control set model predictive control to control the power converter using pulse-width modulation. The solution requires small computation efforts and operates utilizing the assumption that optimal current reference profiles for each torque reference and angular position can be evaluated offline from the magnetization surface of the electrical machine. Thus, knowing the current reference and magnetization surface, the voltage commands for the PWM-driven inverter can be evaluated using simple lookup tables.

One aspect also addressed in this issue was the materials used in the electrical machines. Thus, H. Tiismus et al. presented [7] an additively manufactured soft magnetic transformer core. This was the first time that an electromagnetic device with a fully 3D-printed magnetic core was evaluated in terms of efficiency and performance.

An algorithm to remove the DC drift from the *B-H* curve of an additively manufactured soft ferromagnetic material, based on the sliding mean value subtraction from each cycle of calculated magnetic flux density (*B*) signal, is presented by Bilal Asad et al. in [8]. This is crucial for the accurate estimation of iron losses, and the measurements taken at different flux density values show the effectiveness of the proposed method, whose benefits are presented in the paper.

M. Sato et al. [9] propose a motor in which a composite ring made from a resin material mixed with magnetic powder is mounted on the stator to suppress spatial harmonics. This is a significant method to improve the efficiency of the motor at high speed.

3. Future Developments

Considerable research and investment have been focused on this area. However, despite all of the completed work, there are still many challenges to be faced. One important challenge that needs significant research, development, and investment is related to the integration of electrical machines into existing systems and the new ones that are constantly emerging. The appearance of new materials allows contributions to a greater optimization of this process, which is permanent, and it is expected that it will be continuously developed. This is a fundamental area for the development of more efficient machines. Another fundamental aspect has been the modeling of electrical machines. Nevertheless, there is still much work that needs to be undertaken in this context. Finally, the optimal design of electrical machines in the context of a special application is another aspect that requires further research and study.

4. Conclusions

The articles presented in this Special Issue cover important aspects of electrical machine design and optimization. There are many topics related to electrical machines, but this

Special Issue intends to provide a contribution to the magnetic material modeling of electrical machines to stimulate the community and develop current research, furthering its progress. Therefore, from our perspective, the presented papers will have practical importance for the forthcoming developments in the field of electrical machines.

Author Contributions: All authors contributed equally to this work. All authors have read and agreed to the published version of the manuscript.

Funding: This research was funded by national funds through FCT-Fundação para a Ciência e a Tecnologia, under projects UIDB/50021/2020 and UIDB/00066/2020.

Acknowledgments: The authors are grateful to the MDPI Publisher for the invitation to act as guest editors of this Special Issue and are indebted to the editorial staff of "Energies" for the kind cooperation, patience and committed engagement. We would also like to thank the staff and reviewers for their efforts and professional work.

Conflicts of Interest: The authors declare no conflict of interest.

References

1. Garbiec, T.; Jagiela, M. Accounting for Magnetic Saturation Effects in Complex Multi-harmonic Model of Induction Machine. *Energies* **2020**, *13*, 4670. [CrossRef]
2. Balasubramanian, A.; Martin, F.; Billah, M.M.; Osemwinyen, O.; Belahcen, A. Application of Surrogate Optimization Routine with Clustering Technique for Optimal Design of an Induction Motor. *Energies* **2021**, *14*, 5042. [CrossRef]
3. Garcia-Amorós, J.; Marín-Genescà, M.; Andrada, P.; Martínez-Piera, E. Two-Phase Linear Hybrid Reluctance Actuator with Low Detent Force. *Energies* **2020**, *13*, 5162. [CrossRef]
4. Cao, J.; Guo, X.; Fu, W.; Wang, R.; Liu, Y.; Lin, L. A Method to Improve Torque Density in a Flux-Switching Permanent Magnet Machine. *Energies* **2020**, *13*, 5308. [CrossRef]
5. Pires, V.F.; Foito, D.; Cordeiro, A.; Chaves, M.; Pires, A.J. PV Generator-Fed Water Pumping System Based on a SRM with a Multilevel Fault-Tolerant Converter. *Energies* **2022**, *15*, 720. [CrossRef]
6. Anuchin, A.; Demidova, G.L.; Hao, C.; Zharkov, A.; Bogdanov, A.; Šmídl, V. Continuous Control Set Model Predictive Control of a Switch Reluctance Drive Using Lookup Tables. *Energies* **2020**, *13*, 3317. [CrossRef]
7. Tiismus, H.; Kallaste, A.; Belahcen, A.; Rassolkin, A.; Vaimann, T.; Shams Ghahfarokhi, P. Additive Manufacturing and Performance of E-Type Transformer Core. *Energies* **2021**, *14*, 3278. [CrossRef]
8. Asad, B.; Tiismus, H.; Vaimann, T.; Belahcen, A.; Kallaste, A.; Rassõlkin, A.; Ghafarokhi, P.S. Sliding Mean Value Subtraction-Based DC Drift Correction of B-H Curve for 3D-Printed Magnetic Materials. *Energies* **2021**, *14*, 284. [CrossRef]
9. Sato, M.; Takazawa, K.; Horiuchi, M.; Masuda, R.; Yoshida, R.; Nirei, M.; Bu, Y.; Mizuno, T. Reducing Rotor Temperature Rise in Concentrated Winding Motor by Using Magnetic Powder Mixed Resin Ring. *Energies* **2020**, *13*, 6721. [CrossRef]

Article

Accounting for Magnetic Saturation Effects in Complex Multi-harmonic Model of Induction Machine

Tomasz Garbiec * and Mariusz Jagiela

The Faculty of Electrical Engineering, Automatic Control and Informatics, Opole University of Technology, 45–758 Opole, Poland; m.jagiela@po.edu.pl
* Correspondence: t.garbiec@po.edu.pl

Received: 4 August 2020; Accepted: 7 September 2020; Published: 8 September 2020

Abstract: Computations of quasi-dynamic electromagnetic field of induction machines using the complex magnetic vector potential require the use of the so-called effective magnetization curves, i.e., such in which the magnetic permeability is proportional to the amplitudes of magnetic flux density B or magnetic field strength H, not their instantaneous values. There are several definitions of that parameter mentioned in the literature provided for the case when B or H are monoharmonic. In this paper, seven different methods of determining the effective magnetization curves are compared in relation to the use of a field-circuit multi-harmonic model of an induction machine. The accuracy of each method was assessed by computing the performance characteristics of a solid-rotor induction machine. One new definition of the effective permeability was also introduced, being a function of multiple variables dependent on amplitudes of all the harmonics considered. The analyses demonstrated that the best practical approach, even for the multi-harmonic case, is to express the effective magnetic permeability as the ratio of the amplitudes of the fundamental harmonics of the magnetic flux density and the magnetic field strength, and assuming the sinusoidal variation of the latter.

Keywords: magnetic permeability; effective parameters; induction motor; finite element method

---

1. Introduction

Numerical models based on the use of the finite element method have become an indispensable tool in the design of electrical machines. As is well known, the most accurate approach for overall consideration of physical phenomena in an analyzed converter is the use of time-domain models. Unfortunately, this approach usually involves significant computational cost, sometimes making its practical application impossible. For that reason, it is justified to explore methods which allow, particularly in the analysis of steady states, to achieve similar results, but in a much shorter time. While analyzing induction machines, one of such methods is the use of the multi-harmonic field-circuit model [1–9]. The results presented in [1,2,9] prove that this type of the field-circuit model can be particularly useful when analyzing steady states of the special constructions of induction machines (single-phase machines, high-speed solid-rotor machines). As demonstrated by Garbiec et al., with an appropriate modification of the coupling scheme between individual sub-models, the multi-harmonic field-circuit model makes it possible to achieve comparable calculation results with those obtained using the time-domain model, at a calculation cost comparable to the monoharmonic model. The multi-harmonic model in this case is understood as the De-Gersem type model formulated using the magnetic vector potential, where permeance slot harmonics of the magnetic field distribution in the air gap are extracted via Fourier transform of an air-gap magnetic field. These harmonics can affect the power losses in the rotor and the electromagnetic torque developed by the machine.

The magnetic field in the stator in these models is considered as a function of the fundamental time harmonic only.

The application of the above-mentioned approach essentially involves the problem of modeling the nonlinearity of the magnetic materials. Taking this into account in the monoharmonic field models based on the use of the complex magnetic vector potential method is not a new task and it has been described to date in several works, including [10–16]. They propose several different methods of defining the so-called effective magnetic permeability based, among others, on the development of nonlinear waveforms in the Fourier series [14,16], averaging the magnetic reluctivity [10,15,16] or equivalence of energy stored in ferromagnetic components [10–13]. However, analyzing the conclusions drawn in the mentioned works, it is not possible to clearly state which method of defining the effective magnetic permeability will be most appropriate when using the multi-harmonic model with a strong coupling, as proposed by Garbiec et al. Some of them assume the sinusoidal variation of the magnetic flux density in calculating the effective magnetic permeability (which facilitates the calculations using the magnetic vector potential) [12–15], while others assume that it should be determined assuming sinusoidal variation of the magnetic field strength, or that it is insignificant for the accuracy of the calculations [10,11,14]. The latter approach was adopted in the authors' previous work, and satisfactory results were achieved [9]. It was based on the results of the earlier work of the authors, where the effective magnetic permeability defined with the assumption of the sinusoidal variation of the magnetic field strength provided correct results, modeling a solid-rotor induction machine with the use of the multi-harmonic field-circuit model with a weak coupling [17]. To the best of authors' knowledge, the research providing an analysis of the accuracy of definitions of the effective magnetic permeability on the results of computations using multi-harmonic field-circuit models was not reported.

This paper is a continuation of works by Garbiec et al. that aim at an in-depth verification of the De-Gersem type multi-harmonic complex model of induction machines from the point of view of their application in designing and optimization of real machines. The authors focus on the analysis of the impact of formulating the effective magnetic permeability. For that purpose, the basic operating characteristics of an exemplary machine were calculated for seven different methods of formulating the effective magnetic permeability. The first six methods use existing definitions derived in literature. The available formulas use averaging of magnetic reluctivity in terms of mean or RMS value. Also, the Fourier series truncated to the fundamental component and assuming sinusoidal variation of the magnetic flux density or magnetic field strength is used [14–16]. The seventh method considered in this paper is a new approach proposed by the authors that relies upon formulating the so-called multidimensional effective magnetic permeability. The results of calculations were compared with those obtained with the time-domain model and with the results of measurements performed on the physical model under one of the previous works where the efficient computational method for fast determination of the performance characteristics of solid-rotor induction machines was presented [17].

2. Mathematical Model

2.1. Effective Magnetic Permeability

Definition of the effective magnetic permeability, as used in the previous works of the authors [9,14,16,17], is the reference point for the considerations presented herein:

$$\mu_{\mathrm{I}}\left(H_{pk}\right) = \frac{2}{\pi H_{pk}} \int_0^{\pi} \left(\mu_{\mathrm{DC}}\left(H_{pk}\sin\alpha\right)H_{pk}\sin\alpha\right)\sin\alpha\, d\alpha, \tag{1}$$

where H_{pk}—amplitude of the sinusoidal magnetic field strength H, and μ_{DC}—DC magnetic permeability. The above definition corresponds to the ratio of the fundamental harmonic amplitude of the magnetic flux density waveform (obtained on the assumption of the sinusoidal variation of the magnetic field strength) and H_{pk}. This approach requires a successive determination of the magnetic field strength in a given iteration of the algorithm based on the knowledge of its value determined in the previous iteration.

Assuming the sinusoidal variation of the magnetic field strength, two further definitions can be introduced [16]:

$$\mu_{\mathrm{II}}\left(H_{pk}\right) = \frac{\pi}{\int_0^\pi v_{DC}\left(H_{pk}\sin\alpha\right)d\alpha}, \tag{2}$$

$$\mu_{\mathrm{III}}\left(H_{pk}\right) = \frac{1}{\sqrt{\frac{1}{\pi}\int_0^\pi v_{DC}^2\left(H_{pk}\sin\alpha\right)d\alpha}}, \tag{3}$$

where $v_{DC} = \frac{1}{\mu_{DC}}$. The expressions (2) and (3) correspond to the inverse of the mean value and the inverse of the RMS value of the magnetic reluctivity obtained for the half of the period.

Similarly, the definitions of the effective magnetic permeability can be derived, assuming the sinusoidal variation of the magnetic flux density B with the amplitude B_{pk} [15,16]:

$$\mu_{\mathrm{IV}}\left(B_{pk}\right) = \frac{\pi B_{pk}}{2\int_0^\pi\left(v_{DC}\left(B_{pk}\sin\alpha\right)B_{pk}\sin\alpha\right)\sin\alpha\,d\alpha}, \tag{4}$$

$$\mu_{\mathrm{V}}\left(B_{pk}\right) = \frac{\pi}{\int_0^\pi v_{DC}\left(B_{pk}\sin\alpha\right)d\alpha}, \tag{5}$$

$$\mu_{\mathrm{VI}}\left(B_{pk}\right) = \frac{1}{\sqrt{\frac{1}{\pi}\int_0^\pi v_{DC}^2\left(B_{pk}\sin\alpha\right)d\alpha}}. \tag{6}$$

Each of those definitions assumes that the magnetic saturation coefficient of the ferromagnetic material is a function of the fundamental harmonic of the magnetic field only, expressed by B or H. The definitions (1), (3), (5) and (6) are taken from the literature, whereas (2) and (4) are the equivalents of (5) and (1), respectively.

2.2. Analyzed Machine

In this work, the effect of how the effective permeability is taken into account was examined on the example of a high-speed induction machine with a uniform solid rotor analyzed previously by Garbiec et al. The basic data of the considered motor is presented in Figure 1 and Table 1.

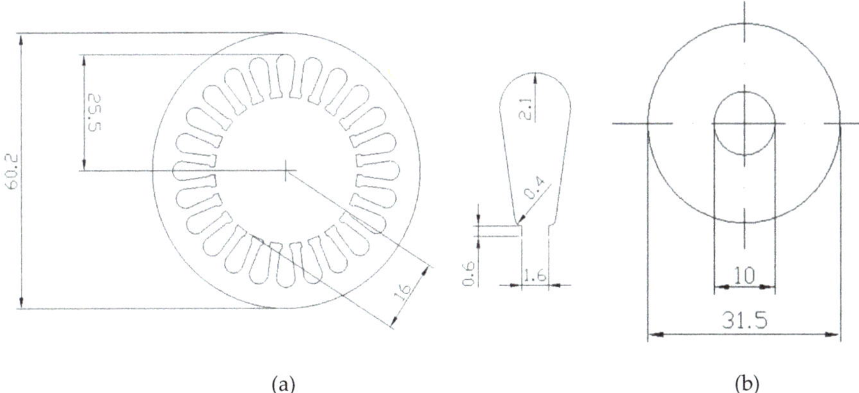

(a)   (b)

Figure 1. Solid-rotor induction machine taken into consideration: (**a**) dimensions of the stator package (cross section area), (**b**) dimensions of the rotor.

Table 1. Basic specifications of tested induction motor.

Parameter	Value
Operation frequency (ω)	2000π rad/s
Number of pole pairs (p)	2
RMS phase voltage	50 V
Phase resistance	0.62 Ω
End-winding leakage	98 μH
Rotor conductivity	5.2 MS/m
Machine length (l_z)	32 mm
Number of stator slots	24

A field-circuit multi-harmonic model with a strong coupling was developed for the analyzed machine. Based on the results of the analysis presented in by Garbiec et al., it was decided to reduce the number of the rotor models to three, i.e., those related to the fundamental harmonic of the magnetic field in the air gap and two most influential higher slot harmonics with ordinal numbers—11 and 13. Neglecting the rotor-end effects the system of equations describing the model takes the form [9]:

$$
\begin{bmatrix}
\mathbf{M}_{11}(\mu_{eff}) & \mathbf{M}_{12} & \mathbf{M}_{13} \\
\mathbf{M}_{21} & \mathbf{M}_{22} & 0 \\
\mathbf{M}_{31} & 0 & 0
\end{bmatrix}
\begin{bmatrix}
\overline{\varphi} \\
\overline{\mathbf{I}} \\
\overline{\lambda}
\end{bmatrix}
=
\begin{bmatrix}
0 \\
\mathbf{E} \\
0
\end{bmatrix},
\tag{7}
$$

where: $\mathbf{M}_{11}$—matrix describing the magnetic and electrical properties of materials, $\mathbf{M}_{12} = -\mathbf{M}_{21}^{T}/(j\omega l_z)$—matrix describing the distribution and connection method of the stator winding, $\mathbf{M}_{13} = \mathbf{M}_{31}^{T}$—matrices describing coupling between the rotor models and the stator model, $\mathbf{M}_{22}$—stator winding impedance matrix, φ—vector of the nodal values of the complex magnetic vector potential, $\mathbf{I}$—vector of the amplitudes of the loop currents in the stator winding, λ—vector of complex circulations of the magnetic field strength vector, $\mathbf{E}$—vector of the complex voltage amplitudes in the loops in the stator winding circuit.

For a comparison, in this paper, calculations were also carried out using a developed complementary time-domain model which uses the DC magnetization curves. The equations for such a model after discretization via Galerkin procedure and the implicit Euler method take the form:

$$
\begin{bmatrix}
\mathbf{S}(\mu_{DC}) + \mathbf{G}\Delta t^{-1} & -\mathbf{D}^{T}\mathbf{K}^{T} \\
l_z\mathbf{K}\mathbf{D}\Delta t^{-1} & \mathbf{K}(\mathbf{R} + \mathbf{L}\Delta t^{-1})\mathbf{K}^{T}
\end{bmatrix}^{n}
\begin{bmatrix}
\varphi \\
\mathbf{i}
\end{bmatrix}^{n}
=
\begin{bmatrix}
\mathbf{G}\Delta t^{-1} & 0 \\
l_z\mathbf{K}\mathbf{D}\Delta t^{-1} & \mathbf{K}\mathbf{L}\Delta t^{-1}\mathbf{K}^{T}
\end{bmatrix}^{n-1}
\begin{bmatrix}
\varphi \\
\mathbf{i}
\end{bmatrix}^{n-1}
+
\begin{bmatrix}
0 \\
\mathbf{K}\mathbf{e}
\end{bmatrix}^{n-1},
\tag{8}
$$

where: $\mathbf{S}$—reluctivity matrix, $\mathbf{G}$—conductivity matrix, $\mathbf{D}$—matrix describing the winding, $\mathbf{K}$—matrix describing the winding connection method, $\mathbf{R}$—winding resistance matrix, $\mathbf{L}$—winding leakage inductivity matrix, Δt—integration step, φ—vector of nodal values of the vector magnetic potential, $\mathbf{i}$—vector of instantaneous values of the stator currents, $\mathbf{e}$—vector of the instantaneous supply voltages values. The rotational movement was modeled using the moving band technique. The computing cost for the multi-harmonic model (when choosing an appropriate coupling scheme) is of the order of a per cent compared to that using the time-domain model [9]. This is the main reason for developing the model described by the Equation (7), however, in this paper the authors focus on the accuracy rather than on the computing cost. The nonlinearity in both above formulated computational problems is solved iteratively by updating, respectively μ_{eff} and μ_{DC} from iteration to iteration unless the relative change of solution is below a prescribed value.

Based on the DC magnetization curves for the stator and rotor materials, the effective magnetization curves were determined according to (1)–(6) (Figure 2). For saturated material, each effective magnetization curve is noticeably characterized by higher magnetic flux density values compared to the characteristic for the direct current. At a low level of magnetic saturation, the effective curves ale very close to the DC curve. Furthermore, a relatively small difference is found for different curves determined assuming the sinusoidal variation of the magnetic flux density for the rotor material. Moreover, the curves calculated with the assumption of a sinusoidal variation of the magnetic flux density are closer to the DC curve than the case with the assumption of a sinusoidal variation of the magnetic field strength. Additionally, curves (4)–(6) are closer to each other than (1)–(3). This is due to the completely different shape of the magnetic flux density waveforms (magnetic field strength) in the case when the magnetic field strength (magnetic flux density) is sinusoidal.

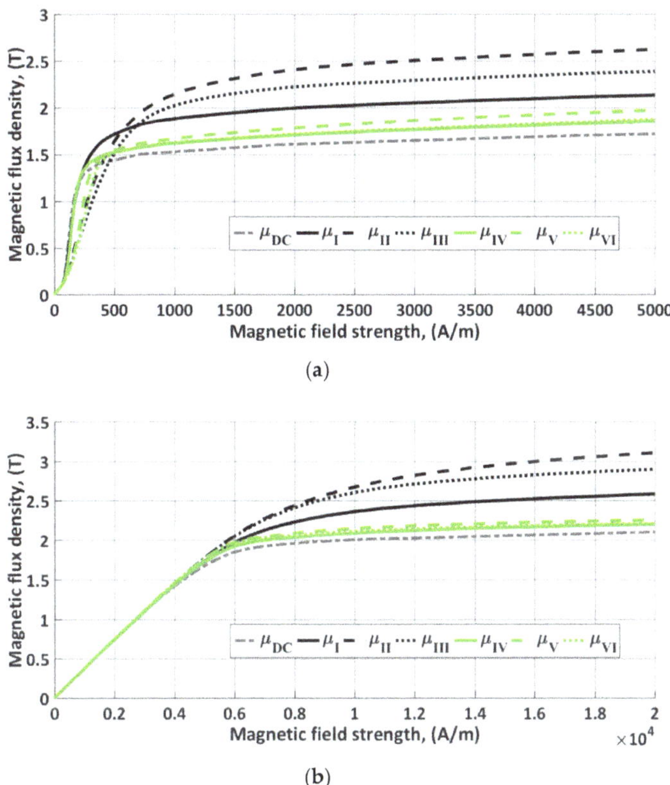

(a)

(b)

Figure 2. Effective magnetization curves determined from the expressions (1)–(6): (a) stator, (b) rotor.

3. Calculation Results

The magnetization curves shown in Figure 2 were used to determine three basic operational characteristics of the machine (electromagnetic torque, RMS phase current, power losses in the rotor) vs. rotational speed at two different supply voltages (phase voltages), i.e., 50 V (rated) and 75 V (150% of the rated value) without changing the frequency of power supply. Note that the effective magnetic permeability for the rotor models is determined in each iteration based on the knowledge of the magnetic field strength distribution (expressions (1)–(3)) or magnetic flux density ((4)–(6)) only for the rotor model associated with the fundamental harmonic of the magnetic field distribution. The characteristics calculated were compared with the results of calculations using time-domain model, as described in Section 2. It should be kept in mind that the considered multi-harmonic model, despite the associated mathematical burden, is a simplified approach. This simplification can be observed in e.g., the obtained steady-state distributions of the magnetic flux over the machine cross-section shown in (Figure 3). Based on the results of that comparison, two groups of characteristics were created

(Figure 4) showing the total relative percentage calculation error for the three determined quantities (electromagnetic torque, RMS phase current, power loss in the rotor) in relation to the calculation results obtained from the model (8):

$$\varepsilon = \frac{\varepsilon_T + \varepsilon_I + \varepsilon_P}{3} \times 100\%, \tag{9}$$

where: $\varepsilon_T = \left|\frac{T_t - T_p}{T_t}\right|$, $\varepsilon_I = \left|\frac{I_t - I_p}{I_t}\right|$, $\varepsilon_P = \left|\frac{P_t - P_p}{P_t}\right|$, T_t, I_t, P_t—respectively, electromagnetic torque, stator RMS phase current, power loss in the rotor calculated with the model formulated in the time domain, T_p, I_p, P_p—respectively, electromagnetic torque, stator RMS phase current, and power loss in the rotor calculated using the multi-harmonic model. Table 2 shows the averaged values of that indicator against the entire speed range considered.

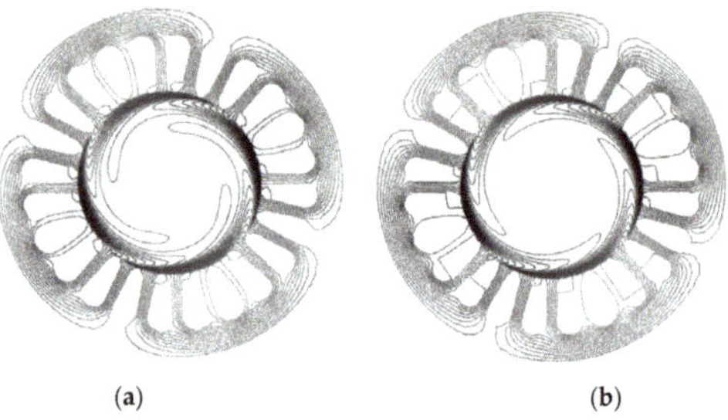

(a)   (b)

Figure 3. Comparison of sample distributions of magnetic flux over cross section of the motor at rotor slip equal to 10%: (**a**) time-domain model, (**b**) multi-harmonic model with all harmonics considered (the magnetic field in the rotor area is a sum of the complex magnetic vector potential distributions obtained from solution of (7) for the individual rotor submodels).

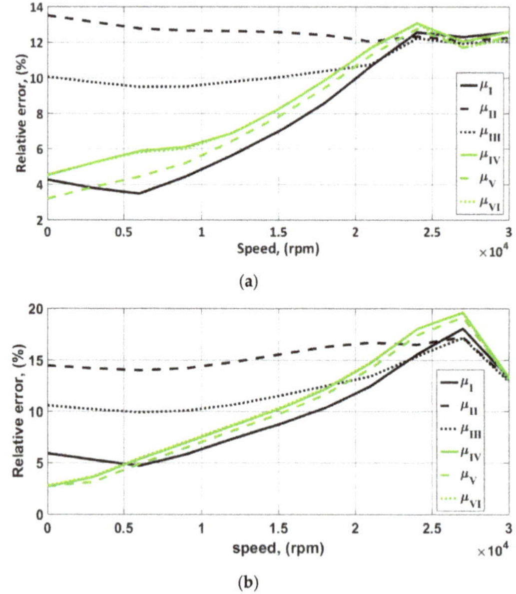

Figure 4. Variation of relative error (9) vs. rotational speed using various methods of calculating the effective magnetic permeability for the RMS phase supply equal to: 50 V (**a**), and 75 V (**b**).

Table 2. Averages percent value of the relative error for $N = 11$ considered points of the curves.

Parameter	μ_{I}	μ_{II}	μ_{III}	μ_{IV}	μ_{V}	μ_{VI}
$\frac{\sum_{i=1}^{N} \varepsilon_i}{N}$ for 50 V	7.75	12.57	10.54	8.72	8.03	8.64
$\frac{\sum_{i=1}^{N} \varepsilon_i}{N}$ for 75 V	9.75	15.17	12.18	10.46	10.04	10.49

Out of the definitions of the effective magnetic permeability considered, noticeably the best results were achieved using μ_{I} (originally used in calculations). Slightly worse results (except for the locked rotor state) can be obtained with the definitions assuming the sinusoidal variation of the magnetic flux density. On the other hand, an interesting conclusion is that the method how the nonlinearities are taken into account has practically no impact on the results of calculations for near-synchronous rotational speeds. It is caused by a much lower saturation of the rotor surface, crucial from the point of view of the machine properties, and consequently use by the time-harmonic model (7) similar values of effective permeability (see Figure 2). An increased error in the range of nominal operation point (21–27 krpm) is an adverse trend observed. However, note that the waveforms of the global error indicator shown in Figure 4 provide information on the discrepancy in the results of calculations of power losses in the rotor (ε_P). The results of those calculations are distinguished by the greatest discrepancy (for μ_I and $N = 11$ considered points of the curve $\frac{\sum_{i=1}^{N} \varepsilon_{Pi}}{N} = 15.42\%$ at 50 V and 19.32% at 75 V), thus overestimate the global error indicator. In the rotational speed range from 21,000 rpm to 27,000 rpm, the average electromagnetic torque error and the RMS current error is 7.56% and 1.43% at 50 V, 10.93% and 1.67% at 75 V, respectively.

Calculating the effective magnetic permeability based on the distribution of the magnetic field strength or magnetic flux density, determined only for the model associated with the fundamental field harmonic in the machine air gap, can raise doubts. To take into account the impact of the higher harmonics and to assess the error related to this assumption, a new definition of the multidimensional effective magnetic permeability was proposed:

$$\mu_{VII}\left(H_{1pk}, H_{11pk}, H_{13pk}\right) = \frac{\sqrt{B_{1pk}^2 + B_{11pk}^2 + B_{13pk}^2}}{\sqrt{H_{1pk}^2 + H_{11pk}^2 + H_{13pk}^2}}, \tag{10}$$

where $H_{1pk}, H_{11pk}, H_{13pk}$ are the magnetic field strength amplitudes related to the fundamental, eleventh and thirteenth harmonics of the field and

$$B_{npk} = \tfrac{2}{\pi} \int_0^{\pi} \mu_{DC}\left(H_{1pk} \sin \alpha + H_{11pk} \sin 11\alpha + H_{13pk} \sin 13\alpha\right)$$
$$\left(H_{1pk} \sin \alpha + H_{11pk} \sin 11\alpha + + H_{13pk} \sin 13\alpha\right) \sin n\alpha \, d\alpha \, \text{for} \, n = 1, 11, 13 \tag{11}$$

For zero values of the higher harmonics, the definition (10) is equivalent to (1). This approach only slightly complicates the model implementation (in each algorithm iteration the distribution of the magnetic field strength in all considered rotor models needs to be obtained). An advantage of this expression, however, is the fact that effective permeability can be expressed before starting the calculations, without knowing the individual amplitudes. For the expression (10), it is possible since B_{npk} does not depend on the phase angles of individual harmonics. In general, this occurs when the frequencies of the higher harmonics are the multiples of the fundamental harmonic. The value of the multidimensional effective permeability is determined only once as a multidimensional table for the assumed variability ranges of individual harmonics of the magnetic field strength. Figure 5 shows the determined distribution of this quantity assuming that $H_{13pk} = 0$ A/m. The multidimensional permeability determined was used again for calculations and compared with the results obtained for (1). The result of that comparison is shown in Table 3. As can be seen, the application of the multidimensional effective permeability brings only slight improvement to the accuracy of calculation results. This is due to relatively low magnetic field strength values in the areas of the models associated with the slot harmonics, as illustrated in Figure 6.

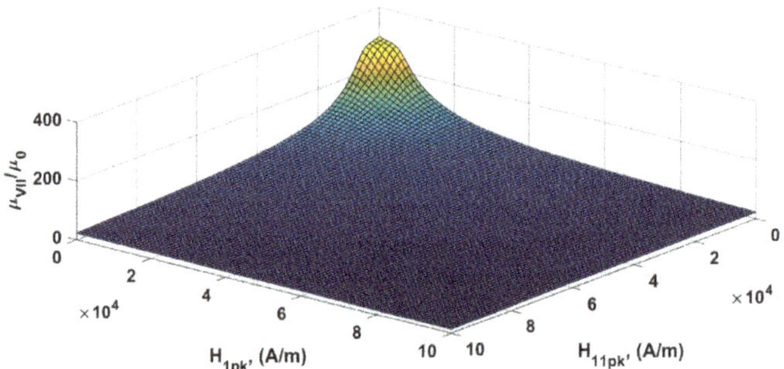

Figure 5. Relative multidimensional effective permeability calculated for the DC magnetization curve of the rotor, assuming zero value of the magnetic field strength of the thirteenth harmonic.

Table 3. Averages percent value of the relative error for $N = 11$ considered points of the curves.

Parameter	μ_{I}	μ_{VII}
$\frac{\sum_{i=1}^{N} \varepsilon_i}{N}$ for 50 V	7.75	7.62
$\frac{\sum_{i=1}^{N} \varepsilon_i}{N}$ for 75 V	9.75	9.59

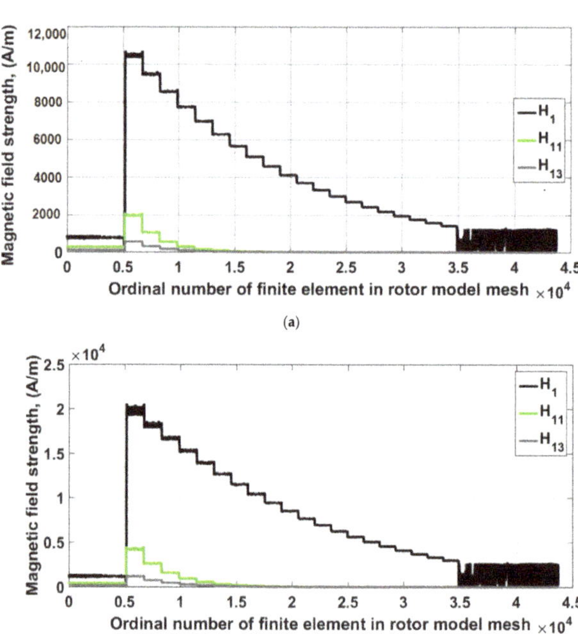

Figure 6. Comparison of the magnetic field strength values in the individual elements creating FE rotor mesh for the rotor models associated with the considered harmonics for a 20% slip: (**a**) at 50 V, (**b**) at 75 V.

4. Physical Validation

The analysis demonstrated that, from a practical point of view, it is sufficient to assume the first definition of the effective magnetic permeability, both due to the accuracy of calculations, and ease of implementation. It should be mentioned that, when adopting the sinusoidal variation of the magnetic field strength in determining the effective magnetic permeability, the fixed point method is applied in the multi-harmonic model to solve the nonlinearity problem, which is easy to implement, and it is distinguished by good convergence quality in case of time-harmonic magnetic fields. In the latter case, it is recommended to use the Newton–Raphson method [16].

The model developed, together with the first definition of the effective magnetic permeability, was applied to calculate the RMS current of a real machine (supplied via a quasi-square wave voltage inverter, fundamental frequency of supplying voltage equal to 500 Hz) with three different solid rotors tested by Garbiec and to compare the results with the calculations performed using a model with a weak coupling the same definition of the effective magnetic permeability applied (see Figure 7).

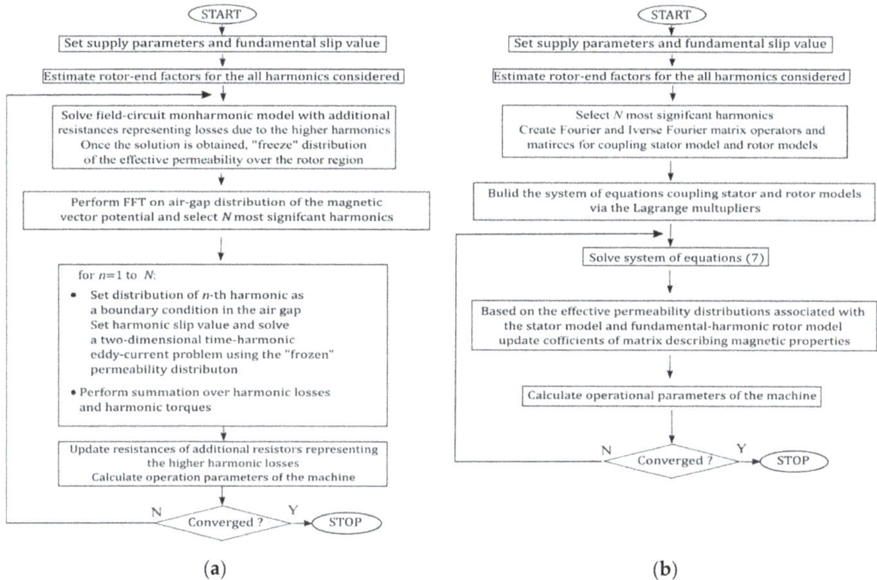

(a) (b)

Figure 7. Comparison of the ideas of the different multi-harmonic model: (**a**) model with the weak coupling, (**b**) model with the strong coupling.

As it can be seen in the figure the idea of using the model with the strong coupling is to simultaneously solve the problem composed of a few submodels (the stator model and rotor models associated with the considered harmonics) coupled by appropriate boundary conditions imposed on the interface boundary in the air gap. This formulation results in one system of equations and determination of the magnetic vector potential distribution for all models via single solution of (7). The interface conditions in the air-gap are formulated with the use of matrix operators of the Fourier transform and the inverse Fourier transform, as shown in Garbiec at. al, enforce the propagation of only the selected harmonics of the air-gap field distribution between the stator model and the considered rotor models. All rotor models act via the same constraint on the stator model. In the considered model with the weak coupling, the monoharmonic field-circuit model is used. The power loss due to slot harmonics is estimated by means of extraction of these harmonics from the calculated distribution of the magnetic vector potential. In this way it is possible to solve the field problem limited to only the area of rotor. In this approach the power factor, torque and stator current can be appropriately corrected, although there is no backward relationship, i.e., the air-gap field remains unaffected by these effects. To this end, it was also necessary to take into account the rotor-end effect [18].

The physical test-stand and result of the comparison of computed and measured data for the machine with different rotors is shown in Figure 8. With the application of the multi-harmonic model with the strong coupling and with the modified sub-model coupling scheme, it was possible to achieve practically identical calculation results, yet in a shorter time by approx. 30%. The errors found result mainly from supplying the physical model via a quasi-square wave voltage inverter (in numerical models, the pure sine-wave voltage supply with an RMS value equal to the RMS value of the measured voltage is assumed).

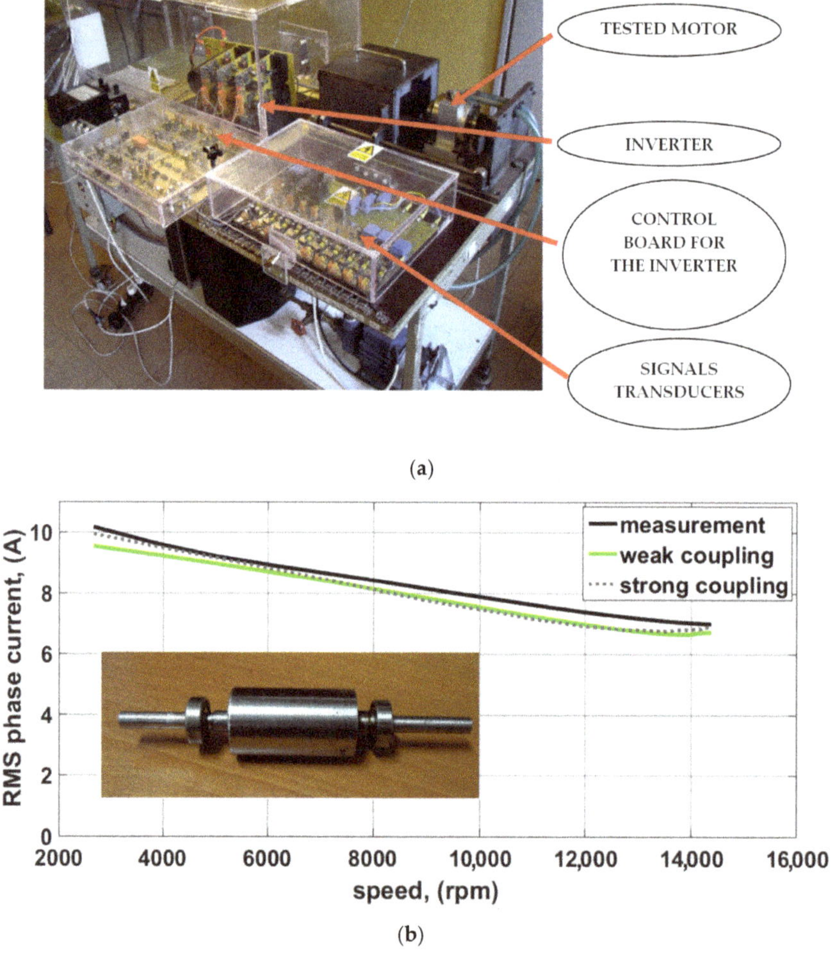

Figure 8. *Cont.*

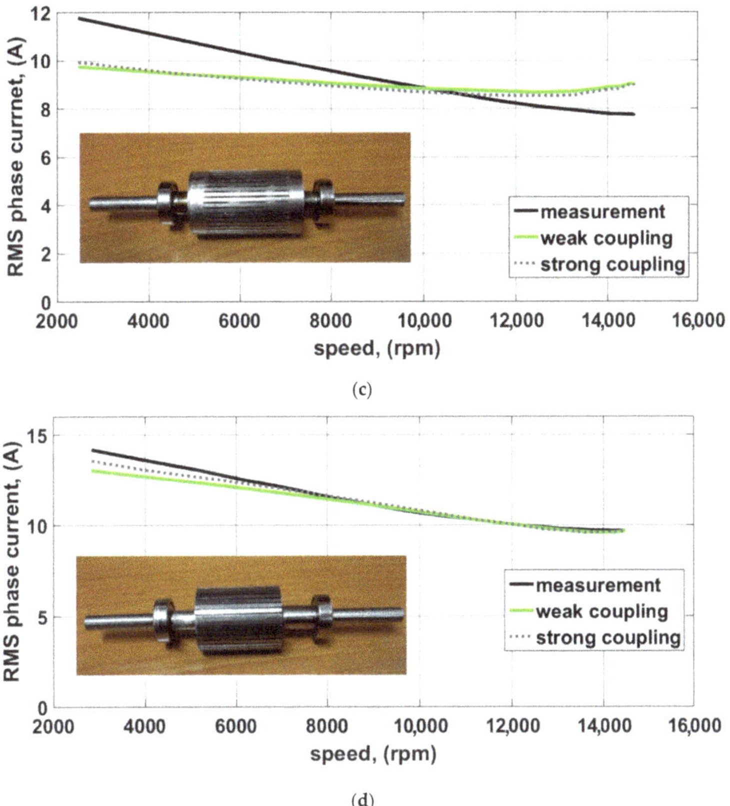

(c)

(d)

Figure 8. Comparison of calculation results with the multi-harmonic model with the weak coupling to those of calculations with the multi-harmonic model with the strong coupling, with the measurement results [17]. (**a**) Physical test-stand, (**b**) RMS current vs. rotor speed for the machine equipped with an uniform rotor, (**c**) RMS current vs. rotor speed for the machine equipped with a slitted rotor with uniform end regions, (**d**) RMS current vs. rotor for a machine equipped with a slitted rotor with slits through the whole length.

5. Conclusions

The comparison of various methods for calculating the effective magnetic permeability for a multi-harmonic model of a solid-rotor induction machine carried out in the study proved that, from a practical point of view, the best results are obtained using the basic definition, described by Equation (1). Determining the effective magnetic permeability as the ratio of the magnitudes of the fundamental harmonics of the magnetic flux density and the magnetic field strength calculated assuming the sinusoidal variation of the latter allows for an easy implementation of the algorithm which is important from practical point of view. Furthermore, it was demonstrated that, in the studied case, the most influential magnetic saturation effects come from the fundamental harmonic of the magnetic field which means that the slot harmonics of the air-gap magnetic flux density do not have to be computed in that process. For the same definition of the effective magnetic permeability and an appropriate modification of the coupling scheme, it is possible to obtain practically the same results as in the case of the multi-harmonic model with the weak coupling, but yet in a shorter time. Another step to take will be to conduct detailed research using the developed model to analyze the magnetic saturation effects in the squirrel cage induction machines with both three- and single-phase windings.

Author Contributions: Conceptualization, T.G. and M.J.; methodology, T.G. and M.J.; software, T.G.; investigation, T.G.; resources, T.G.; data curation, T.G.; writing—original draft preparation, T.G.; writing—review and editing, M.J.; visualization, T.G.; supervision, M.J.; project administration, T.G. All authors have read and agreed to the published version of the manuscript.

Funding: This research received no external funding.

Conflicts of Interest: The authors declare no conflict of interest.

References

1. De Gersem, H.; Hameyer, K. Air-gap flux splitting for the time-harmonic finite-element simulation of single-phase induction machines. *IEEE Trans. Magn.* **2002**, *38*, 1221–1224. [CrossRef]
2. De Gersem, H.; De Brabandere, K.; Belmans, R.; Hameyer, K. Motional time-harmonic simulation of slotted single-phase induction machines. *IEEE Trans. Energy Convers.* **2002**, *17*, 313–318. [CrossRef]
3. Räisänen, V.; Suuriniemi, S.; Kurz, S.; Kettunen, L. Rapid Computation of Harmonic Eddy-Current Losses in High-Speed Solid-Rotor Induction Machines. *IEEE Trans. Energy Convers.* **2013**, *28*, 782–790. [CrossRef]
4. Räisänen, V. Air Gap Fields in Electrical Machines: Harmonics and Modeling of Movement. Ph.D. Thesis, Tampere University of Technology, Tampere, Finland, 2015.
5. Räisänen, V.; Suuriniemi, S.; Kettunen, L. Generalized Slip Transformations and Air-Gap Harmonics in Field Models of Electrical Machines. *IEEE Trans. Magn.* **2016**, *52*. [CrossRef]
6. Mezani, S.; Laporte, B.; Takorabet, N. Saturation and space harmonics in the complex finite element computation of induction motors. *IEEE Trans. Magn.* **2005**, *41*, 1460–1463. [CrossRef]
7. Ouazir, Y.; Takorabet, N.; Ibtiouen, R.; Touhami, O.; Mezani, S. Consideration of space harmonics in complex finite element analysis of induction motors with an air-gap interface coupling. *IEEE Trans. Magn.* **2006**, *42*, 1279–1282. [CrossRef]
8. Rainer, S.; Biro, O.; Stermecki, A.; Weilharter, B. Frequency Domain Evaluation of Transient Finite Element Simulations of Induction Machines. *IEEE Trans. Magn.* **2012**, *48*, 851–854. [CrossRef]
9. Garbiec, T.; Jagiela, M.; Kulik, M. Application of Nonlinear Complex Polyharmonic Finite-Element Models of High-Speed Solid-Rotor Induction Motors. *IEEE Trans. Magn.* **2020**, *56*. [CrossRef]
10. Paoli, G.; Bíró, O.; Buchgraber, G. Complex representation in nonlinear time harmonic eddy current problems. *IEEE Trans. Magn.* **1998**, *34*, 2625–2628. [CrossRef]
11. Vassent, E.; Meunier, G.; Sabonnadier, J.C. Simulation of induction machine operation using complex magnetodynamic finite elements. *IEEE Trans. Magn.* **1989**, *25*, 3064–3066. [CrossRef]
12. Hedia, H.; Remacle, J.-F.; Dular, P.; Nicolet, A.; Genon, A.; Legros, W. A sinusoidal magnetic field computation in nonlinear materials. *IEEE Trans. Magn.* **1995**, *31*, 3527–3529. [CrossRef]
13. Stermecki, A.; Biro, O.; Preis, K.; Rainer, S.; Ofner, G. Numerical analysis of steady-state operation of three-phase induction machines by an approximate frequency domain technique1. *E I Elektrotech. Inftech.* **2011**, *128*, 81–85. [CrossRef]
14. Takorabet, N.; Laporte, B.; Mezani, S. An approach to compute saturated induction motors in steady state. In Proceedings of the IEEE International Electric Machines and Drives Conference, Madison, WI, USA, 1–4 June 2003; Volume 2, pp. 1646–1650. [CrossRef]
15. Du Terrail, Y.; Sabonnadiere, J.C.; Masse, P.; Coulomb, J.L. Nonlinear complex finite elements analysis of electromagnetic field in steady-state AC devices. *IEEE Trans. Magn.* **1984**, *20*, 549–552. [CrossRef]
16. Arkkio, A. Analysis of Induction Motor Based on the Numerical Solution of the Magnetic Field and Circuit Equations. Ph.D. Thesis, Helsinki University of Technology, Espoo, Finland, 1987.
17. Garbiec, T. Fast Computation of Performance Characteristics for Solid-Rotor Induction Motors with Electrically Inhomogeneous Rotors. *IEEE Trans. Energy Convers.* **2016**, *31*, 1688–1696. [CrossRef]
18. Pyrhonen, J.; Nerg, J.; Kurronen, P.; Lauber, U. High-Speed High-Output Solid-Rotor Induction-Motor Technology for Gas Compression. *IEEE Trans. Ind. Electron.* **2009**, *57*, 272–280. [CrossRef]

Article

Application of Surrogate Optimization Routine with Clustering Technique for Optimal Design of an Induction Motor

Aswin Balasubramanian *, Floran Martin *, Md Masum Billah, Osaruyi Osemwinyen and Anouar Belahcen *

Department of Electrical Engineering and Automation, Aalto University, 02150 Espoo, Finland;
md.billah@aalto.fi (M.M.B.); osaruyi.osemwinyen@aalto.fi (O.O.)
* Correspondence: aswin.balasubramanian@aalto.fi (A.B.); floran.martin@aalto.fi (F.M.);
anouar.belahcen@aalto.fi (A.B.)

Abstract: This paper proposes a new surrogate optimization routine for optimal design of a direct on line (DOL) squirrel cage induction motor. The geometry of the motor is optimized to maximize its electromagnetic efficiency while respecting the constraints, such as output power and power factor. The routine uses the methodologies of Latin-hypercube sampling, a clustering technique and a Box–Behnken design for improving the accuracy of the surrogate model while efficiently utilizing the computational resources. The global search-based particle swarm optimization (PSO) algorithm is used for optimizing the surrogate model and the pattern search algorithm is used for fine-tuning the surrogate optimal solution. The proposed surrogate optimization routine achieved an optimal design with an electromagnetic efficiency of 93.90%, for a 7.5 kW motor. To benchmark the performance of the surrogate optimization routine, a comparative analysis was carried out with a direct optimization routine that uses a finite element method (FEM)-based machine model as a cost function.

Keywords: induction motors; surrogate optimization; Box–Behnken design; Latin-hypercube sampling; clustering; particle swarm optimization; pattern search

Citation: Balasubramanian, A.; Martin, F.; Billah, M.M.; Osemwinyen, O.; Belahcen, A. Application of Surrogate Optimization Routine with Clustering Technique for Optimal Design of an Induction Motor. *Energies* **2021**, *14*, 5042. https://doi.org/10.3390/en14165042

Academic Editor: Mario Marchesoni

Received: 8 July 2021
Accepted: 13 August 2021
Published: 17 August 2021

Publisher's Note: MDPI stays neutral with regard to jurisdictional claims in published maps and institutional affiliations.

1. Introduction

Electrical machines have a wide range of use cases, from household utilities to industrial applications, which consume a huge share of all the generated electrical energy [1]. To reduce the global greenhouse gas emissions, it is important to design electrical machines with high energy efficiency. The characteristics of the electrical machines are usually analyzed with a finite element method (FEM)-based electromagnetic simulation for better accuracy. The output characteristics of the electrical machine are highly sensitive to the design variables and the global search optimization algorithms, such as particle swarm optimization (PSO) or genetic algorithm (GA), require many model evaluations to reach the desired optimal solution [2,3]. This causes the optimization process with a FEM-based machine model as a cost function to be computationally expensive [4]. To utilize the time and computational resources efficiently, surrogate optimization techniques are used to optimize the electrical machines, which requires only a few FEM simulations for evaluation.

Response surface methodology (RSM) is a technique used to develop a polynomial function for a complex FEM-based multi-physics model, which defines the relationship between design variables and the output response of an electrical machine [5–9]. This polynomial model can be used with an optimization algorithm to search for the optimal solution. The Box–Behnken design, one of the popular response surface approaches, is used in conjunction with the FEM-based machine model to generate second-order polynomial functions for objective and constraints of the electrical machine [10–14]. The range of the boundaries of the design variables affect the accuracy of the polynomial function and in turn the optimal solution of the electrical machine.

In this article, a novel surrogate optimization routine is proposed for optimizing a three-phase direct on line (DOL) squirrel cage induction motor. The rotor bars of the induction machine are not skewed for the sake of the demonstration of the surrogate optimization routine. The aim of the new routine is to discretize the problem domain into a number of subdomains for improving the accuracy of the polynomial models for a better search of the optimal solution, while efficiently using the computational resources with a smaller number of FEM simulations. To utilize the full capacity of the computational resource, the routine is programmed in such a way as to handle 15 FEM simulations in parallel. The methodologies used in the optimization process are Latin-hypercube sampling for design of the experiments, a clustering algorithm for dividing the problem domain and a Box–Behnken design as a response surface approach. The particle swarm optimization (PSO) algorithm is used for optimizing the polynomial functions of the response surfaces, while a pattern search algorithm is used for fine-tuning the surrogate optimal solution from the PSO. For the purpose of visualization, the proposed surrogate optimization routine is demonstrated with a simplification of the design problem, leaving three design variables. Validation of the results from the proposed surrogate optimization routine for a multivariate design problem is performed by comparing it to the direct optimization routine, which uses FEM simulation as a cost function.

2. Optimization Problem

The electrical machine analyzed in the optimization problem is a three-phase squirrel cage induction motor for a direct online industrial application. The electrical steel core material used in the motor is M400-50A and the rotor cage is made of aluminum. The goal of the optimization problem is to maximize the electromagnetic efficiency, η, satisfying the constraints of the output power, P_{out}, and power factor, PF, for a given volume of the machine. The outer diameter of the stator, D_{se}, and axial length of the machine, l, are fixed so that the volume of the machine remains constant throughout the optimization process. The rotor end-ring overhang length, l_{oh}, is kept constant so that the cross-section area of the end ring depends only on the height of the rotor slot. The specifications and fixed parameters of the induction motor for the optimization problem are shown in Table 1. The objective and constraints of the optimization problem are specified in Table 2. The optimization variables and their ranges for the induction motor are shown in Figure 1 and Table 3. The analysis of the machine design is done with timestepping simulation of a 2D finite element solver software, FCSMEK, developed by the research group of electromechanics at Aalto university [15]. The simulation of the timestepping analysis computes the electromagnetic characteristics of electrical machines by solving the circuit and field equations with the Crank–Nicholson timestepping method. The time is discretized at short time intervals and the magnetic field, currents, and potentials of the windings are solved at successive instants of time. The rotation of the rotor is accomplished by changing the finite element mesh in the air gap. The non-linear system of equations obtained at each timestep is solved using the Newton–Raphson method. The core losses are evaluated using the modified Jordan loss equation with a two-component loss model, namely with eddy current loss and hysteresis loss. The excess losses are included in the dynamic eddy current loss computation [15]. For simplicity, the mechanical losses and other manufacturing losses are not considered for comparing the results of the optimal solutions. Hence, the electromagnetic efficiency η is computed as shown in Equation (1).

$$\eta = \frac{P_{in} - P_{elec}}{P_{in}} \times 100\% \tag{1}$$

where P_{in} is the input power of the induction motor and P_{elec} is the electromagnetic loss of the induction machine, which is comprised of iron losses and copper losses of the stator and rotor. The steady state temperature of the stator and rotor are considered as 80 °C and 100 °C, respectively, for the simulation.

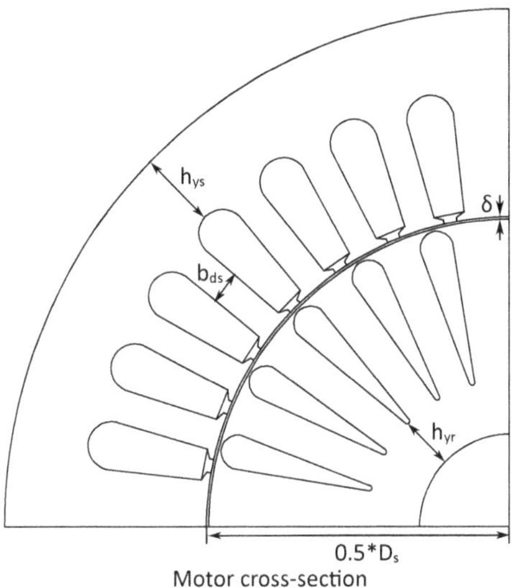
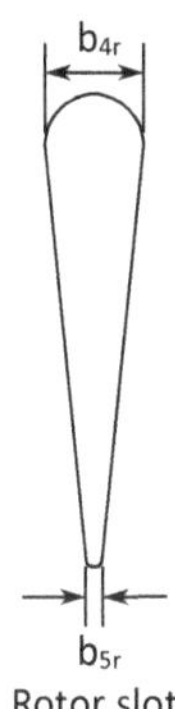

Figure 1. Optimization variables of the induction motor.

Table 1. Specifications and fixed parameters of the induction motor.

Parameter	Value
Output power, P_{out} [W]	7500
Line voltage, U_{l} [V]	400
Frequency, f [Hz]	50
Number of poles, p	4
Filling factor-stator slot, K_{Cu}	0.6
Number of conductors-stator slots, Z_{Qs}	28
Number of parallel paths-stator windings, a	2
Number of stator slots, N_{s}	48
Number of rotor slots, N_{r}	44
Axial length of the machine, l (mm)	220
Stator outer diameter, D_{se} (mm)	227.7
Rotor inner (shaft) diameter, D_{ye} (mm)	48.98
End-ring overhang length, l_{oh} (mm)	40
Conductivity—aluminum (20 °C), σ_{Al} [S/m]	35.5×10^{6}
Conductivity—copper (20 °C), σ_{Cu} [S/m]	57×10^{6}

Table 2. Objective and constraints of the optimization problem.

Objective	Goal
Electromagnetic efficiency, η (%)	To maximize the electromagnetic efficiency, η
Constraints	**Range**
Output power, P_{out} [W]	$7500 \leq P_{\mathrm{out}} \leq 7600$
Power factor, PF	$PF \geq 0.78$

Table 3. Optimization variables and their range.

Optimization Variables	Range
Stator inner diameter, D_s (mm)	$120 \leq D_s \leq 150$
Stator tooth width, b_{ds} (mm)	$2 \leq b_{ds} \leq 4$
Stator yoke width, h_{ys} (mm)	$10 \leq h_{ys} \leq 30$
Slip, s (%)	$1 \leq s \leq 2.1$
Air gap width, δ (mm)	$0.4 \leq \delta \leq 0.7$
Rotor slot upper width, b_{4r} (mm)	$3 \leq b_{4r} \leq 6$
Rotor slot lower width, b_{5r} (mm)	$0.5 \leq b_{5r} \leq 2$
Rotor yoke width, h_{yr} (mm)	$2 \leq h_{yr} \leq 15$

3. Response Surface Methodology Optimization with Box–Behnken Design

Response surface methodology (RSM) is a set of mathematical and statistical techniques used to draw a relationship between control variables (inputs) and output response. This relationship can be approximated into a polynomial model, which can be useful in predicting the response of the control variables, hypothesis testing and finding the optimal condition of the variable settings [16]. In practice, the response surface methodology can be applied to simulate experimental results or for constructing a surrogate function of a computationally expensive multi-physics model. Optimizing the geometric variables of the induction motor directly with the finite element model is computationally expensive. In this article, for boundary-constrained input variables the RSM is used for approximating the electrical quantities of an induction machine into a second-order polynomial model. The second-order polynomial model that describes the functional relationship of the RSM between the control variables and the output response is as shown in Equation (2) [17].

$$y = \beta_0 + \sum_{i=1}^{k} \beta_i x_i + \sum_{i=1}^{k} \beta_{ii} x_i^2 + \sum_{i=1}^{k-1} \sum_{j=i+1}^{k} \beta_{ij} x_i x_j \tag{2}$$

where y is the output response, x_i and x_j are the input control variables, $\beta_0, \beta_i, \beta_{ii}, and \beta_{ij}$ are the coefficients of the input control variable terms, and k is the number of control variables. The coefficients are estimated as shown in Equation (3).

$$\beta = [X^T X]^{-1} X^T Y \tag{3}$$

where X is the matrix of input control variables sampled at multiple points and Y is the corresponding output response vector. The surrogate function shown in Equation (2) works well for interpolation of design variables to predict the output response, but prediction of output response by extrapolation of design variables can be inaccurate. One of the commonly used designs for determining the response surface, as shown in Equation (2), is the Box–Behnken design codeveloped by Box and Behnken in 1960 [18]. If the control variable space is defined as a cube, then the sample points are taken at the geometric center of the cube and at the middle points of the edges of the cube. The Box–Behnken design sample points represented for a three-dimensional control variable space are shown in Figure 2. Positioning sample points in this way preserves a uniform variance within the definition of the hyper-cube [18]. The number of sample points, N, for a given number of control variables is shown in Equation (4) [19].

$$N = 2k(k-1) + C_0 \tag{4}$$

where k is the number of design variables and C_0 is the number of center points.

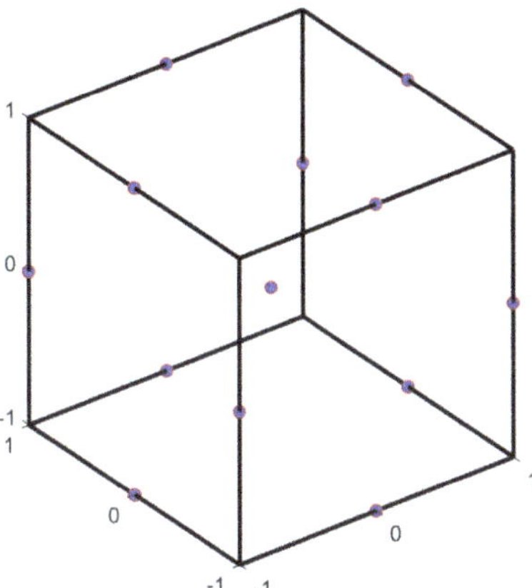

Figure 2. Box–Behnken design represented for a 3-dimensional variable space.

The Box–Behnken design was applied to the problem of an induction machine as described in Section 2 for performing the surrogate optimization. The process flow of optimization with the Box–Behnken design is presented in Figure 3. For easier representation, the optimization variables presented in Table 3 are assumed as $x_1, x_2, x_3 \ldots, x_8$ in their respective order. Based on the optimization variable boundaries from Table 3, the Box–Behnken design sample points were created for the variables as an eight-dimensional hypercube. These samples were simulated with FCSMEK finite element software for calculating their corresponding response characteristics, such as efficiency, η, output power, P_{out}, and power factor, PF. The relationship between the optimization variables and the output response was established as a second-order polynomial function as shown in Equation (2). The coefficients of the polynomial terms were calculated from the predetermined output responses by FEM simulations sampled at Box–Behnken sample points as shown in Equation (3). The polynomial response functions for the surrogate optimization problem are presented in Appendix A. These surrogate functions were used as the cost function of the PSO algorithm. A population of 1000 particles of the PSO was initialized with the Latin-hypercube sampling method.

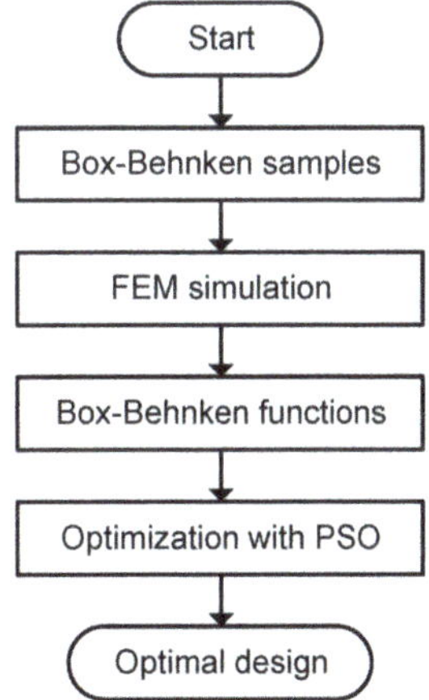

Figure 3. Flow chart: optimization with Box–Behnken design.

The objective and constraints of the problem are as shown in Table 2. The optimal motor design from Box–Behnken design is validated with FEM simulation as shown in

Table 4. It is observed that a difference in the result of the objective electromagnetic efficiency, η, between the surrogate optima and its FEM validation is considerably high. Moreover, the optimal solution does not respect the constraints of the output power, P_{out}, and the power factor, PF, coupled with a high margin of error. The accuracy of the surrogate functions is impacted by the application of the Box–Behnken design to a large design variable space. Hence, a new optimization routine is proposed in this article (Section 6) for improving the accuracy of the surrogate functions resulting in an improved optimal solution.

Table 4. FEM validation of the optimal solution from Box–Behnken design

Output Response	Box–Behnken Design	FEM Validation	Difference
Electromagnetic efficiency, η	95.30%	93.31%	2.13%
Output power, P_{out}	7500.28 W	7131.98 W	5.18%
Power factor, PF	0.7860	0.7267	8.16%

4. Latin-Hypercube Sampling

Latin-hypercube sampling is a statistical method for selecting near-random samples from the input variable space, proposed by McKay, Beckman, and Conover [20]. It uses a sampling scheme of stratification to improve the distribution of samples in the input variable space. To select n Latin-hypercube samples for a sampling function with $x_i = (x_1, x_2)$ as input variables, the range of each of the x_i is stratified into n equiprobable intervals. One observation is selected at random from each of the n intervals. These observations corresponding to x_1 and x_2 are matched at random to form n Latin-hypercube samples. A set of five samples generated with the Latin-hypercube sampling method for the input variables of $x_i = (x_1, x_2)$ is shown in Figure 4. The Latin-hypercube sampling method is widely used in the design of experiments for various applications of computer modeling [21–23]. In this article, the Latin-hypercube sampling method is used in the proposed optimization routine for selecting samples that satisfy a set of criteria and for initializing the first swarm of the PSO algorithm.

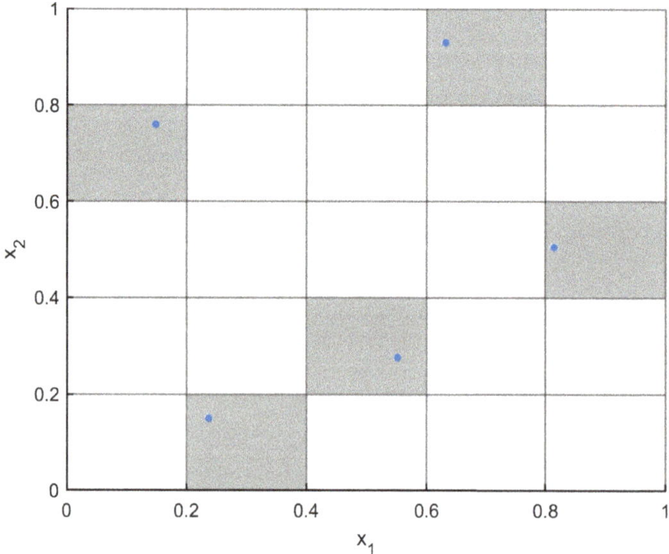

Figure 4. Latin-hypercube sampling.

5. Clustering

Clustering is a method involved in partitioning a given dataset into different groups or clusters. The data that are mapped to a particular cluster tend to have similar characteristics and follow a similar pattern [24]. Clustering helps in classifying and analyzing large datasets, which can be applied in fields of machine learning, data science, pattern recognition, image processing, and bioinformatics [25–27]. The k-means is one of the oldest computational techniques used in solving clustering problems, based on the algorithm proposed by Lloyd [28]. If an integer k is chosen for partitioning a dataset into k clusters and n is the number of data points of the dataset, the goal of Lloyd's (k-means) algorithm is to find k centroids so as to minimize the potential function, γ. The potential function γ is a measure of the total squared Euclidean distance between each data point and its closest centroid, as shown in Equation (5).

$$\gamma = \sum_{j=1}^{k} \sum_{i=1}^{m} \left\| x_i + c_j \right\|^2 \tag{5}$$

where m is the number of data points in the jth cluster, x_i and c_j are the data points and centroid of the jth cluster. Since Lloyd's (k-means) algorithm involves selecting the initial k centroids uniformly at random from the dataset, it suffers from inconsistency and accuracy issues. Arthur and Vassilvitskii proposed a randomized seeding technique for selecting the initial centroids of the k-means and combined it with the original k-means algorithm to call it the k-means++ algorithm with improved speed and accuracy [29]. A set of randomly generated samples clustered with the k-means++ algorithm is shown in Figure 5. In the proposed surrogate optimization routine, the k-means++ algorithm is used in partitioning the samples created from design variables into different clusters. In a multi-variable clustering problem, the variables can have varying scales of magnitude and incomparable units. Thus, it is required to apply the feature scaling technique to normalize the data for standardization. Z-score transformation is one of the successful standardization techniques utilized before applying the k-means clustering method to a dataset [30]. Equation (6) is used to estimate the Z-score,

$$x' = \frac{x - \bar{x}}{\sigma_x} \tag{6}$$

where x is the variable vector that needs to be standardized, $\bar{x}$ and σ_x are the mean and standard deviation of the vector x, and x' is the transformed variable vector that implies the z-score.

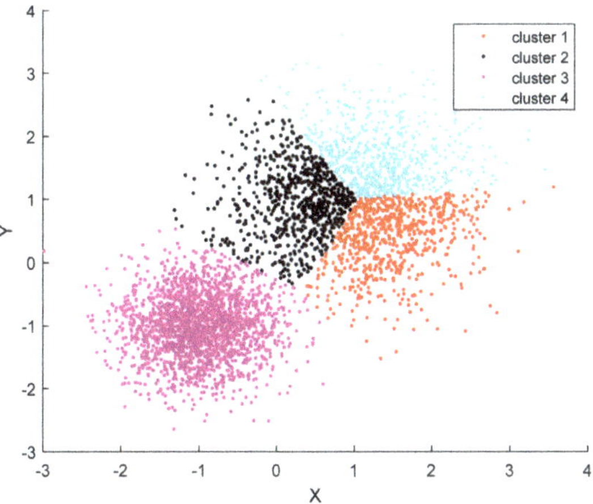

Figure 5. A set of randomly generated samples clustered with the k-means++ algorithm.

6. The Proposed Surrogate Optimization Routine

To overcome the issue in the accuracy of the surrogate model as shown in Section 3, a new surrogate optimization routine is proposed for the induction machine problem. The concepts of Latin-hypercube sampling, clustering, and Box–Behnken design are used in the proposed optimization routine, and algorithms such as PSO and pattern search are used for optimizing the control variables of the problem statement. The proposed optimization routine is divided into two parts, namely a surrogate optimization part and a pattern search part, which are presented as flow charts in Figures 6 and 7, respectively. The implementation of the surrogate optimization routine was carried out with MATLAB programming.

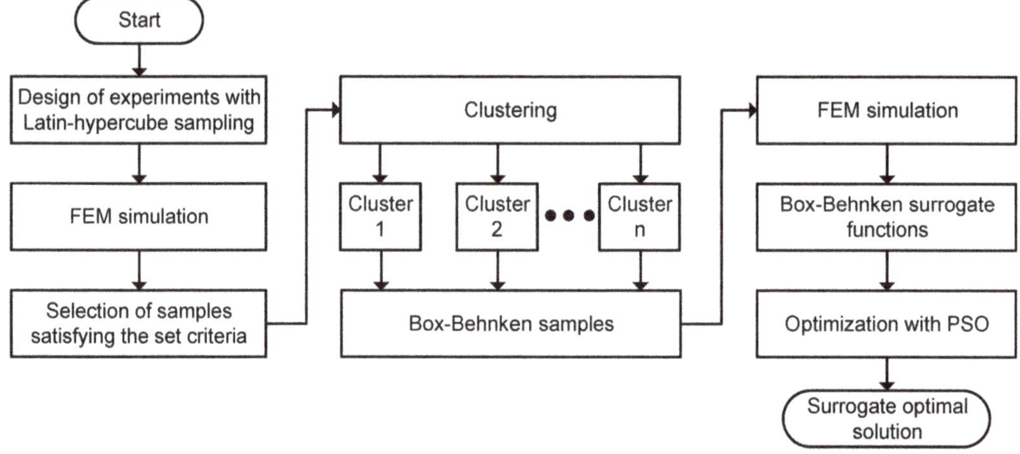

Figure 6. Flow chart: proposed optimization routine—part 1 (surrogate optimization).

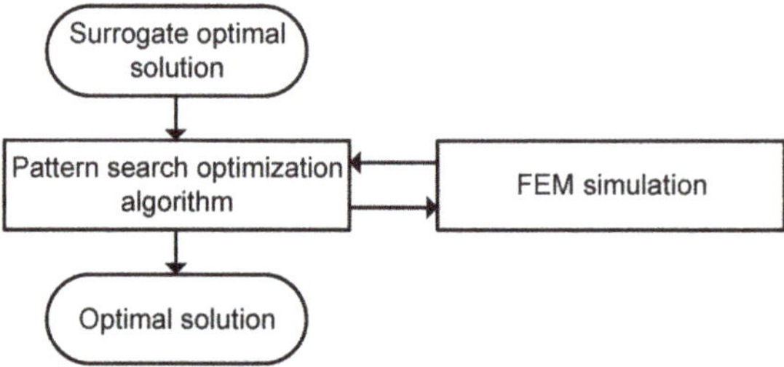

Figure 7. Flow chart: proposed optimization routine—part 2 (pattern search).

The surrogate optimization part begins with the design of experiments using the Latin-hypercube sampling method. The design variables shown in Table 3 were sampled for 500 Latin-hypercube samples within the boundaries of the variable design space. These samples were simulated with FCSMEK finite element software for calculating their output responses, including electromagnetic efficiency, η, output power, P_{out}, and power factor, PF. A set of criteria (satisfying the threshold values of the output responses) was devised to select a set of samples from the design of experiment. The design variables of the selected samples were standardized as shown in Equation (6), and clustered with the k-means ++ algorithm into different groups or clusters. The optimality of the number of clusters was evaluated by the MATLAB built-in function *evalclusters* using the gap statistics criterion [31,32]. The domain of the optimization problem was divided into n clusters and within those clusters, new boundaries of the design variables were established. The surrogate optimization process with Box–Behnken design as explained in Section 3 was applied to each cluster. At the end of the surrogate optimization, n surrogate optimal solutions were obtained. The accuracy of the surrogate optimal solutions was validated with timestepping FEM simulation.

The surrogate optimal solutions were used for initializing the pattern search optimization algorithm. The goal of the pattern search algorithm is to search for an optimal solution in the vicinity of the surrogate optimal solution, which improves the objective of the optimization problem while respecting the constraints. The FEM simulation was used as a cost function for the pattern search algorithm. The final improved optimal solution was obtained at the end of the pattern search optimization process.

7. Results

7.1. Visualization of a 3-Variable Optimization

A three-variable optimization problem for an induction machine is demonstrated to visualize the flow of the proposed surrogate optimization routine as shown in Figure 8. The stator inner diameter, D_s, stator yoke width, h_{ys}, and slip, s, from Table 3 were chosen as the design variables and the remaining variables were fixed to constant values. The objective and constraints of the optimization problem are presented in Table 2. A set of 500 samples of the optimization variables were generated using the Latin-hypercube sampling method as shown in Figure 8a. These samples were simulated with FCSMEK finite element software to calculate the respective output responses. Selection criteria based on the output response of the samples as presented in Table 5 were applied to the Latin-hypercube samples to pick the samples of interest, as shown in Figure 8b. These samples were clustered into different groups as shown in Figure 8c. The Box–Behnken domain of each of the groups was as shown in Figure 8d. The surrogate optimal solution was computed and validated with FEM simulation for each of the clusters. The best solution of the surrogate optimization from one of the clusters is presented in Table 6. It is seen that the difference in the computation of the surrogate optimal solution has been reduced considerably when compared with the results from Table 4, but the constraint of the output power P_{out} is not respected. The pattern search algorithm with FEM simulation as the cost function was applied to the surrogate optimal solution to search for a better solution in its vicinity. The objective and design variables of the optimal solution from the surrogate optimization part and pattern search part are compared in Tables 7 and 8. It can be seen that the pattern search algorithm found a marginally better solution in the neighborhood of the surrogate optimal solution, while respecting the constraints of the optimization problem. The electromagnetic efficiency, power factor and electromagnetic losses are compared for various load points in Tables 9 and 10.

Table 5. Sample selection criteria from the Latin-hypercube sampling method.

Output Response	Sample Selection Criteria
Electromagnetic efficiency, η (%)	$\eta \geq 91$
Output power, P_{out} [W]	$7000 \leq P_{out} \leq 8000$
Power factor, PF	$PF \geq 0.75$

Table 6. Validation of surrogate optimal solution of the proposed optimization routine (part 1—surrogate optimization) with FEM simulation 3-variable optimization problem.

Output Response	Surrogate Optima	FEM Validation	Difference
Electromagnetic efficiency, η	93.63%	93.64%	0.01%
Output power, P_{out}	7599.97 W	7614.76 W	0.19%
Power factor, PF	0.7800	0.7825	0.32%

Table 7. Surrogate optimal solution compared with the improved optimal solution (part 1—surrogate optimization vs. part 2—pattern search) of the proposed optimization routine 3-variable optimization problem.

Output Response	Optimal Solution (Part 1)	Optimal Solution (Part 2)	% Increase
Electromagnetic efficiency, η	93.64%	93.66%	0.021%
Output power, P_{out}	7614.76 W	7572.32 W	−0.56%
Power factor, PF	0.7825	0.7801	−0.31%

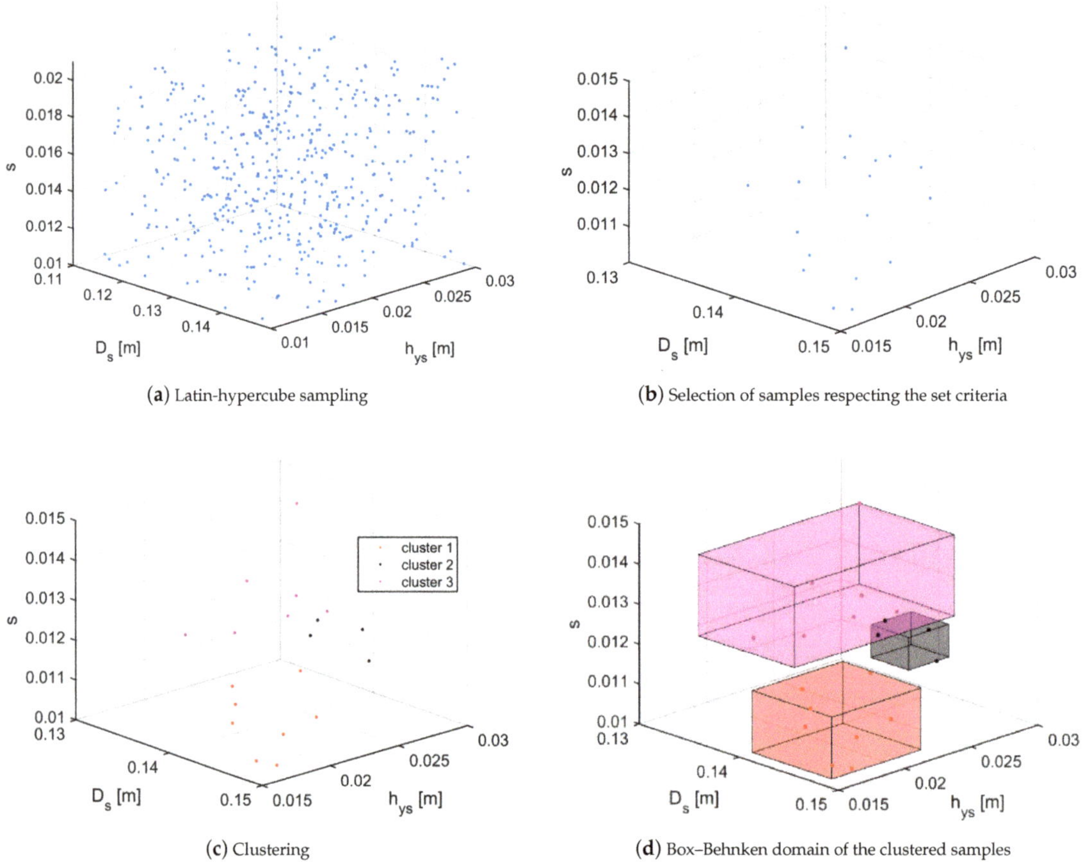

(**a**) Latin-hypercube sampling

(**b**) Selection of samples respecting the set criteria

(**c**) Clustering

(**d**) Box–Behnken domain of the clustered samples

Figure 8. Visualization of the process involved in the proposed surrogate optimization routine.

Table 8. Comparison of design variables of the optimal solution (part 1—surrogate optimization vs. part 2—pattern search) from the proposed surrogate optimization routine 3-variable optimization problem.

Design Variables	Optimal Solution (Part 1)	Optimal Solution (Part 2)
Stator inner diameter, D_s (mm)	147.54	146.65
Stator yoke width, h_{ys} (mm)	17.01	17.27
Slip, s (%)	1.12	1.12

Table 9. Comparison of electromagnetic efficiency and power factor for various load points of the optimal design 3-variable optimization problem.

Output Response	Load (100%)	Load (75%)	Load (50%)
Electromagnetic efficiency, η	93.66%	93.74%	92.97%
Power factor, PF	0.7801	0.7122	0.5888

Table 10. Comparison of losses for various load points of the optimal design 3-variable optimization problem.

Losses	Load (100%)	Load (75%)	Load (50%)
Stator losses, P_{stator}	335.81 W	251.63 W	195.66 W
Rotor losses, P_{rotor}	177.18 W	124.02 W	88.06 W
Total electromagnetic losses, P_{elec}	512.99 W	375.65 W	283.72 W

7.2. Multivariate Optimization

The proposed surrogate optimization routine was applied to a multivariate optimization problem as specified in Section 2. The selection criteria of the samples based on its output response for the clustering process were as shown in Table 5. The output response of the surrogate optimal solution from one of the clusters was validated with FEM simulation as presented in Table 11. It was found that difference in the output responses has been reduced considerably when compared with the solution presented in Table 4. The decrease in difference of the output response leads to the surrogate optimal solution respecting the set of constraints of the optimization problem. The pattern search algorithm improves the objective of the surrogate optimal solution by searching in the vicinity of the surrogate design. The objective and design variables at the end of both the surrogate optimal part and the pattern search part are compared in Tables 12 and 13. It can be noted from Table 13 that to improve the electromagnetic efficiency, η, of the surrogate optimal solution, the values of the design variables, such as air gap width, δ, stator tooth width, b_{ds}, and stator yoke width, h_{ys}, have changed by a small margin. The electromagnetic efficiency, power factor, and electromagnetic losses are compared for various load points in Tables 14 and 15. The flux density distribution of the optimal solution (quadrant of the optimal induction machine) at the end of pattern search algorithm is shown in Figure 9.

Table 11. Validation of surrogate optimal solution of the proposed optimization routine (part 1—surrogate optimization) with FEM simulationdesign–multivariate optimization problem.

Output Response	Surrogate Optima	FEM Validation	Difference
Electromagnetic efficiency, η	93.48%	93.54%	0.064%
Output power, P_{out}	7500.16 W	7514.74 W	0.194%
Power factor, PF	0.7964	0.7970	0.075%

Table 12. Surrogate optimal solution compared with the improved optimal solution (part 1—surrogate optimization vs. part 2—pattern search) of the proposed optimization routine–multivariate optimization problem.

Output Response	Optimal Solution (Part 1)	Optimal Solution (Part 2)	% Increase
Electromagnetic efficiency, η	93.54%	93.90%	0.385%
Output power, P_{out}	7514.74 W	7502.62 W	−0.161%
Power factor, PF	0.7970	0.7803	−2.095%

Table 13. Comparison of design variables of the optimal solution (part 1—surrogate optimization vs. part 2—pattern search) from the the proposed surrogate optimization routine–multivariate optimization problem.

Design Variables	Optimal Solution (Part 1)	Optimal Solution (Part 2)
Stator inner diameter, D_{s} (mm)	142.77	142.77
Stator tooth width, b_{ds} (mm)	3.28	3.30
Stator yoke width, h_{ys} (mm)	19.65	15.80
Slip, s (%)	1.26	1.26
Air gap width, δ (mm)	0.68	0.70
Rotor slot upper width, $b_{4\mathrm{r}}$ (mm)	5.78	5.78
Rotor slot lower width, $b_{5\mathrm{r}}$ (mm)	0.88	0.88
Rotor yoke width, h_{yr} (mm)	6.81	6.81

Table 14. Comparison of electromagnetic efficiency and power factor for various load points of the optimal design–multivariate optimization problem.

Output Response	Load (100%)	Load (75%)	Load (50%)
Electromagnetic efficiency, η	93.90%	93.95%	93.19%
Power factor, PF	0.7803	0.7164	0.5944

Table 15. Comparison of losses for various load points of the optimal design 3-variable optimization problem.

Losses	Load (100%)	Load (75%)	Load (50%)
Stator losses, P_{stator}	305.44 W	234.83 W	185.66 W
Rotor losses, P_{rotor}	181.89 W	127.15 W	88.77 W
Total electromagnetic losses, P_{elec}	487.33 W	361.98 W	274.43 W

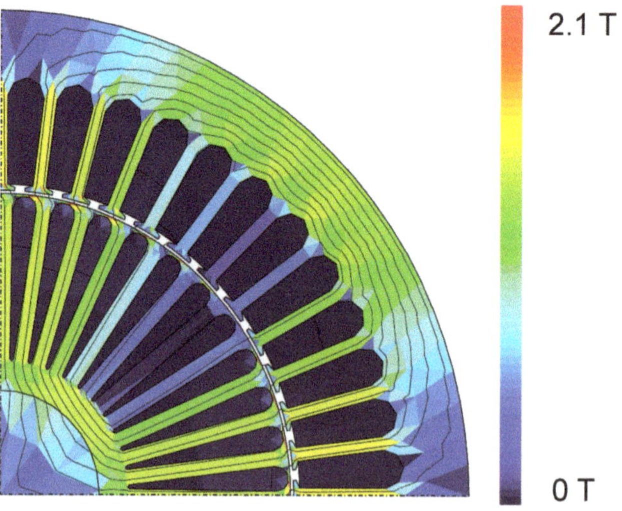

Figure 9. Flux density distribution of the optimal solution from the proposed surrogate optimization routine–multivariate optimization problem.

7.3. Comparison: Proposed Surrogate Optimization Routine vs. Direct Optimization Routine

In this section, the result from the proposed surrogate optimization routine is compared with the direct optimization routine, which uses FEM simulation as the cost function. The flow chart of the direct optimization routine is presented in Figure 10. The computational cost of the direct optimization routine is high since it uses FEM simulation as the cost function. The PSO algorithm and pattern search algorithm were used in the direct optimization routine with the same configuration as that of the proposed surrogate optimization routine. Due to the high computational cost of the FEM simulation and limitation in the computational capabilities of the research workstation, the size of the population was fixed to 30 particles for the PSO algorithm in the direct optimization routine. The optimal solution from the PSO algorithm was used to initialize the pattern search algorithm, which searches for a better solution in the vicinity. The objective and design variables of the optimal solution from the proposed surrogate optimization routine are compared with the optimal solution from the direct optimization routine in Tables 16 and 17. The electromagnetic efficiency, η, of the optimal solution from the proposed surrogate optimization routine reached closer to that of the direct optimization routine. The marginal difference in the design variables slip, s, rotor slot lower width, b_{5r}, and rotor yoke width, h_{yr}, between the routines impacts the electromagnetic efficiency, η, of the optimal solutions. The electromagnetic efficiency, power factor, and electromagnetic losses are compared for various load points in Tables 18 and 19. The advantage of using the proposed surrogate optimization routine is that it requires far fewer FEM simulations than the direct optimization routine, while maintaining an accurate evaluation of the optimal design. The optimization routines were performed in the computer with dual processors of Intel Xeon Silver 4114 CPU at a clock-rate of 2.2 GHz, which can handle parallel computations of 15 FEM simulations. The comparison of the number of FEM simulations in the proposed surrogate optimization routine and direct optimization routine is presented in Table 20.

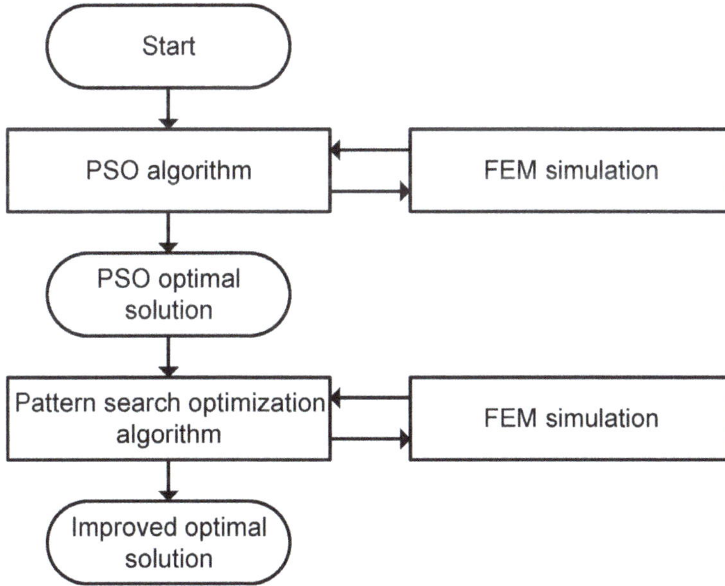

Figure 10. Flow chart: direct optimization routine with FEM as the cost function.

Table 16. Optimal design from the proposed surrogate optimization routine compared with the direct optimization routine.

Output Response	Proposed Surrogate Optimization Routine	Direct Optimization Routine
Electromagnetic efficiency, η	93.90%	93.89%
Output power, P_{out}	7502.62 W	7505.29 W
Power factor, PF	0.7803	0.7801

Table 17. Design variables of the optimal solution from the the proposed surrogate optimization routine compared with direct optimization routine.

Design Variables	Proposed Surrogate Optimization Routine	Direct Optimization Routine
Stator inner diameter, D_{s} (mm)	142.77	142.09
Stator tooth width, b_{ds} (mm)	3.30	3.40
Stator yoke width, h_{ys} (mm)	15.80	15.28
Slip, s (%)	1.26	1.30
Air gap width, δ (mm)	0.70	0.70
Rotor slot upper width, b_{4r} (mm)	5.78	5.80
Rotor slot lower width, b_{5r} (mm)	0.88	1.34
Rotor yoke width, h_{yr} (mm)	6.81	9.50

Table 18. Comparison of electromagnetic efficiency and power factor for various load points of the optimal design—proposed surrogate optimization routine vs. direct optimization routine.

Load	Output Response	Proposed Surrogate Optimization Routine	Direct Optimization Routine
100%	Electromagnetic efficiency, η	93.90%	93.89%
100%	Power factor, PF	0.7803	0.7801
75%	Electromagnetic efficiency, η	93.95%	93.94%
75%	Power factor, PF	0.7164	0.7159
50%	Electromagnetic efficiency, η	93.19%	93.16%
50%	Power factor, PF	0.5944	0.5926

Table 19. Comparison of losses for various load points of the optimal design-proposed surrogate optimization routine vs. direct optimization routine.

Load	Losses	Proposed Surrogate Optimization Routine	Direct Optimization Routine
100%	Stator losses, P_{stator}	305.44 W	302.42 W
100%	Rotor losses, P_{rotor}	181.89 W	185.6 W
100%	Total electromagnetic losses, P_{elec}	487.33 W	488.01 W
75%	Stator losses, P_{stator}	234.83 W	233.65 W
75%	Rotor losses, P_{rotor}	127.15 W	129.37 W
75%	Total electromagnetic losses, P_{elec}	361.98 W	363.03 W
50%	Stator losses, P_{stator}	185.66 W	185.60 W
50%	Rotor losses, P_{rotor}	88.77 W	89.78 W
50%	Total electromagnetic losses, P_{elec}	274.43 W	275.38 W

Table 20. Comparison of the number of FEM simulations in proposed surrogate optimization routine and direct optimization routine.

Parameter	Proposed Surrogate Optimization Routine	Direct Optimization Routine
Number of FEM simulations	1364	75,208

The reliability of the proposed surrogate optimization routine for the induction machine problem was assessed with 20 continuous runs. The electromagnetic efficiency, η, output power, P_{out}, and power factor, PF, of the optimal solution from each run were analyzed to provide the probability distribution as presented in Figure 11. It can be seen that all of the optimal solutions from the proposed surrogate optimization routine respect the constraints specified in the optimization problem and that the range of the objective, electromagnetic efficiency, η, varies between 93.75% and 93.95%.

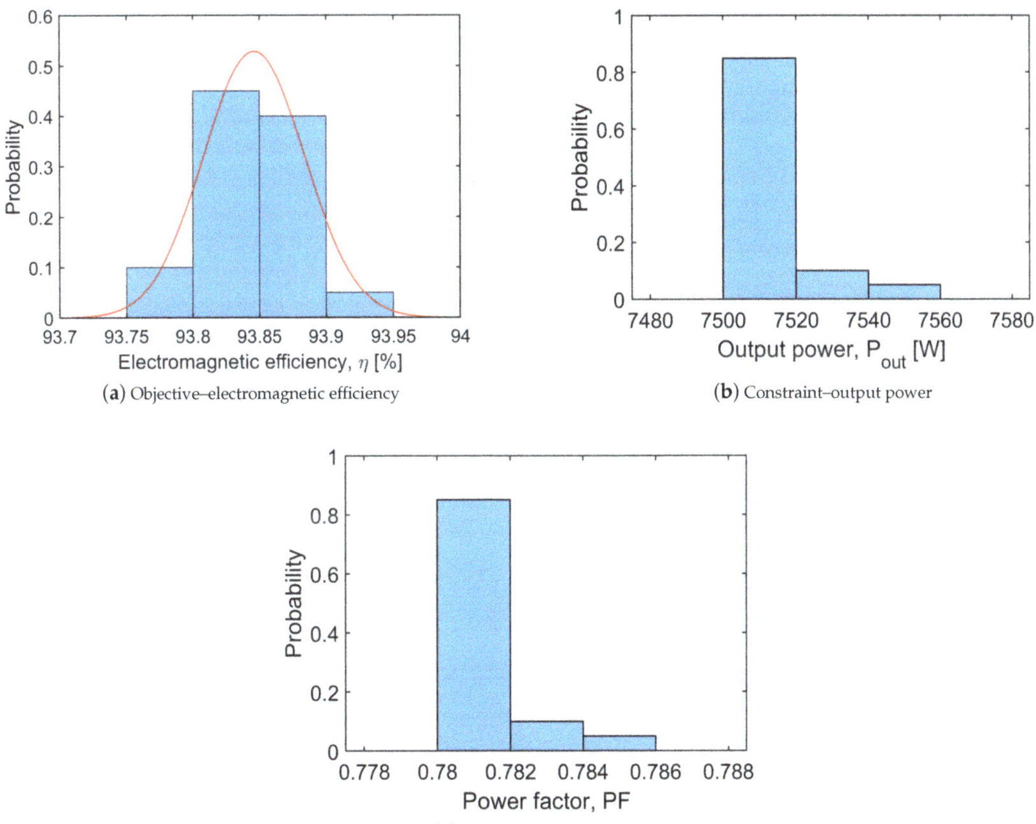

(**a**) Objective–electromagnetic efficiency

(**b**) Constraint–output power

(**c**) Constraint–power factor

Figure 11. Probability distribution of the objective and constraints from 20 runs of the proposed surrogate optimization routine.

8. Conclusions

This paper proposes a novel, efficient, and reliable surrogate optimization routine that can be applied to multiple design problems. The proposed clustering technique used in the routine enables improving the accuracy of the surrogate model while exploring promising subsets of the design variable range. The surrogate optimization routine was applied to design an optimal three-phase induction motor, maximizing its efficiency for a given volume. The surrogate functions of the electromagnetic efficiency, output power, and power factor were constructed as a function of eight design variables and these functions acted as the objective and constraints of the optimization problem. A precision of 0.01 mm was considered for the optimization process, which is possible only with laser cutting of the electrical sheets at the prototyping level. A three-variable optimization problem was performed to demonstrate the discretization of the optimization problem into a few sub-domains with the clustering algorithm for searching for the optimal solution. The results of the proposed surrogate optimization routine applied to the multivariate optimization problem show an improved optimal solution when compared with optimization with a simple Box–Behnken design. This proves the improvement of the accuracy of the surrogate functions by the application of the proposed surrogate optimization routine. To benchmark the proposed surrogate optimization routine, a direct optimization routine was applied to the induction motor problem, which uses FEM simulation as a cost function. Upon comparing the results of both routines, the optimal solution from the proposed surrogate optimization routine was shown to reach closer to that from the direct optimization routine. Additionally, the proposed surrogate optimization routine used 1364 FEM simulations compared with 75,208 FEM simulations of the direct optimization routine, thus greatly improving the computational efficiency. Future work on the proposed surrogate optimization routine will focus on performance evaluations on different types of machines and its application for multi-objective optimization problems with several constraints.

Author Contributions: Conceptualization, A.B. (Aswin Balasubramanian); methodology, A.B. (Aswin Balasubramanian), F.M. and A.B. (Anouar Belahcen); software, A.B. (Aswin Balasubramanian), F.M. and A.B. (Anouar Belahcen); validation, A.B. (Aswin Balasubramanian) and F.M.; writing—original draft preparation, A.B. (Aswin Balasubramanian); writing—review and editing, A.B. (Aswin Balasubramanian), F.M., O.O., M.M.B. and A.B. (Anouar Belahcen); visualization, A.B. (Aswin Balasubramanian) and M.M.B.; supervision, A.B. (Anouar Belahcen); project administration, A.B. (Anouar Belahcen); funding acquisition, A.B. (Anouar Belahcen). All authors have read and agreed to the published version of the manuscript.

Funding: This work was supported in part by the Academy of Finland consortium grant 330747.

Institutional Review Board Statement: Not applicable.

Informed Consent Statement: Not applicable.

Data Availability Statement: Not applicable.

Acknowledgments: We would like to acknowledge our colleague Alireza Nematsaberi, Department of Electrical Engineering and Automation, Aalto University, for his contribution in brainstorming the idea of this article, and Devi Geetha Nair, R&D Engineer, ABB Oy, Finland for proofreading the article.

Conflicts of Interest: The authors declare no conflict of interest.

Abbreviations

The following abbreviations are used in this manuscript:

DOL	Direct On Line
FEM	Finite Element Method
PSO	Particle Swarm Optimization
GA	Genetic Algorithm
RSM	Response Surface Methodology

Appendix A

Table A1. Quadratic closed form equation of Box–Behnken design—constant and first order terms.

Efficiency, η	Output Power, P_{out}	Power Factor, PF	Terms
−2.13	−17,793.93	−7.21	constant
19.70	181,583.20	45.91	x_1
49.97	412,548.73	91.53	x_2
33.81	−253,121.70	114.90	x_3
−66.36	−1,451,699.89	−1295.93	x_4
54.17	−2,003,197.88	11.83	x_5
56.87	−1,835,156.76	176.28	x_6
−14.46	−1749.63	−47.77	x_7
543.42	3,068,544.76	1640.46	x_8

Table A2. Quadratic closed form equation of Box–Behnken design—second-order squared terms.

Efficiency, η	Output Power, P_{out}	Power Factor, PF	Terms
−145.04	−13,854,715.08	−10,487.18	x_1^2
14092.83	203,636,581.33	−49,000.00	x_2^2
−2056.07	−71,109,777.78	−14,800.00	x_3^2
316.60	−27,247,703.79	1164.10	x_4^2
−6796.98	−26,223,650.13	−14,116.67	x_5^2
1444.97	54,209,095.30	12,323.08	x_6^2
−4837.02	300,328.53	−17,066.67	x_7^2
919.65	−492,073.97	3357.69	x_8^2

Table A3. Quadratic closed form equation of Box–Behnken design—second-order product terms.

Efficiency, η	Output Power, P_{out}	Power Factor, PF	Terms
−41.93	−745,736.39	−135.53	$x_1.x_2$
−325.50	−2,094,333.11	−1541.78	$x_1.x_3$
−144.03	−2,518,297.47	−937.01	$x_1.x_4$
−8885.96	358,926,619.68	−172,724.87	$x_1.x_5$
−2609.85	−23,720,999.82	−16,790.21	$x_1.x_6$
−1773.84	−96,582,265.11	−35,390.48	$x_1.x_7$
−81.98	−3,469,143.73	−878.47	$x_1.x_8$
−35253.62	−115,771,295.06	−143,252.98	$x_2.x_3$
−169.52	−1,850,886.93	−4.83	$x_2.x_4$
−101.87	4,251,893.60	−400.60	$x_2.x_5$
250.33	7,026,589.33	7777.78	$x_2.x_6$
80.11	20,428,872.87	1755.56	$x_2.x_7$
−212.22	14,168,262.83	−473.33	$x_2.x_8$
51.70	964,692.96	219.99	$x_3.x_4$

Table A3. *Cont.*

Efficiency, η	Output Power, P_{out}	Power Factor, PF	Terms
-1369.82	$-12{,}319{,}740.72$	-3271.67	$x_3.x_5$
-246.48	$-1{,}619{,}519.09$	-318.64	$x_3.x_6$
397.42	$-2{,}404{,}587.73$	-5716.67	$x_3.x_7$
-550.39	$3{,}492{,}065.36$	-505.00	$x_3.x_8$
-377.48	$9{,}293{,}965.01$	-603.33	$x_4.x_5$
89.05	$-1{,}249{,}662.46$	175.38	$x_4.x_6$
-2335.66	$-22{,}764{,}991.96$	-215.00	$x_4.x_7$
536.32	$-1{,}139{,}470.06$	19757.58	$x_4.x_8$
-575.70	$63{,}834{,}896.15$	-2057.58	$x_5.x_6$
-431.04	$48{,}765{,}232.48$	-1333.33	$x_5.x_7$
83.45	$-10{,}184{,}611.42$	237.06	$x_5.x_8$
-2747.19	$-20{,}314{,}075.20$	-4236.36	$x_6.x_7$
846.01	$-91{,}722{,}135.11$	$45{,}777.77$	$x_6.x_8$
1596.02	$-106{,}614{,}524.44$	$46{,}888.89$	$x_7.x_8$

References

1. Benhaddadi, M.; Olivier, G.; Labrosse, D.; Tetrault, P. Premium efficiency motors and energy saving potential. In Proceedings of the 2009 IEEE International Electric Machines and Drives Conference, Miami, FL, USA, 3–6 May 2009; pp. 1463–1468.
2. Han, W.; Van Dang, C.; Kim, J.W.; Kim, Y.J.; Jung, S.Y. Global-Simplex Optimization Algorithm Applied to FEM-Based Optimal Design of Electric Machine. *IEEE Trans. Magn.* **2017**, *53*, 1–4. [CrossRef]
3. Han, W.; Tran, T.T.; Kim, J.W.; Kim, Y.J.; Jung, S.Y. Mass Ionized Particle Optimization Algorithm Applied to Optimal FEA-Based Design of Electric Machine. *IEEE Trans. Magn.* **2016**, *52*, 1–4. [CrossRef]
4. Slawomir Wiak, P.; Belahcen, A.; Martin, F.; Zaim, M.E.H.; Dlala, E.; Kolondzovski, Z. Combined FE and Particle Swarm algorithm for optimization of high speed PM synchronous machine. *Model. Magn. Electr. Circuits* **2015**, *34*, 475–484. [CrossRef]
5. Jung, J.W.; Lee, B.H.; Kim, K.S.; Kim, S.I. Interior Permanent Magnet Synchronous Motor Design for Eddy Current Loss Reduction in Permanent Magnets to Prevent Irreversible Demagnetization. *Energies* **2020**, *13*, 5082. [CrossRef]
6. Ishikawa, T.; Yamada, M.; Kurita, N. Design of Magnet Arrangement in Interior Permanent Magnet Synchronous Motor by Response Surface Methodology in Consideration of Torque and Vibration. *IEEE Trans. Magn.* **2011**, *47*, 1290–1293. [CrossRef]
7. Lee, J.H. Optimum Shape Design Solution of Flux Switching Motor Using Response Surface Methodology and New Type Winding. *IEEE Trans. Magn.* **2012**, *48*, 1637–1640. [CrossRef]
8. Lee, B.H.; Hong, J.P.; Lee, J.H. Optimum Design Criteria for Maximum Torque and Efficiency of a Line-Start Permanent-Magnet Motor Using Response Surface Methodology and Finite Element Method. *IEEE Trans. Magn.* **2012**, *48*, 863–866. [CrossRef]
9. Semon, A.; Melcescu, L.; Craiu, O.; Crăciunescu, A. Design Optimization of the Rotor of a V-type Interior Permanent Magnet Synchronous Motor using Response Surface Methodology. In Proceedings of the 2019 11th International Symposium on Advanced Topics in Electrical Engineering (ATEE), Bucharest, Romania, 28–30 March 2019; pp. 1–4.
10. Lee, S.; Kim, Y.; Lee, K.; Kim, S. Multiobjective Optimization Design of Small-Scale Wind Power Generator With Outer Rotor Based on Box–Behnken Design. *IEEE Trans. Appl. Supercond.* **2016**, *26*, 1–5. [CrossRef]
11. Han, J.; Lee, J.; Kim, W. A Study on Optimal Design of the Triangle Type Permanent Magnet in IPMSM Rotor by Using the Box–Behnken Design. *IEEE Trans. Magn.* **2015**, *51*, 1–4. [CrossRef]
12. Rafiee, V.; Faiz, J. Robust Design of an Outer Rotor Permanent Magnet Motor Through Six-Sigma Methodology Using Response Surface Surrogate Model. *IEEE Trans. Magn.* **2019**, *55*, 1–10. [CrossRef]
13. Bramerdorfer, G. Computationally Efficient Tolerance Analysis of the Cogging Torque of Brushless PMSMs. *IEEE Trans. Ind. Appl.* **2017**, *53*, 3387–3393. [CrossRef]
14. Zhu, Z.; Zhu, J.; Zhu, H.; Zhu, X.; Yu, Y. Optimization Design of an Axial Split-Phase Bearingless Flywheel Machine with Magnetic Sleeve and Pole-Shoe Tooth by RSM and DE Algorithm. *Energies* **2020**, *13*, 1256. [CrossRef]
15. Arkkio, A. Analysis of Induction Motors Based on the Numerical Solution of the Magnetic Field and Circuit Equations. Ph.D. Thesis, Aalto University, Aalto, Finland, 1987
16. Khuri, A.I.; Mukhopadhyay, S. Response surface methodology. *WIREs Comput. Stat.* **2010**, *2*, 128–149. [CrossRef]

17. Yang, J.; Peterson, J.; Khuri, A.; Goldfarb, H.; Mukhopadhyay, S.; Piepel, G.; Carter, W. *Response Surface Methodology And Related Topics*; World Scientific Publishing Company: Singapore, 2006
18. Box, G.E.P.; Behnken, D.W. Some New Three Level Designs for the Study of Quantitative Variables. *Technometrics* **1960**, *2*, 455–475. [CrossRef]
19. Zolgharnein, J.; Shahmoradi, A.; Ghasemi, J.B. Comparative study of Box–Behnken, central composite, and Doehlert matrix for multivariate optimization of Pb (II) adsorption onto Robinia tree leaves. *J. Chemom.* **2013**, *27*, 12–20. [CrossRef]
20. Mckay, M.D.; Beckman, R.J.; Conover, W.J. A Comparison of Three Methods for Selecting Values of Input Variables in the Analysis of Output From a Computer Code. *Technometrics* **2000**, *42*, 55–61. [CrossRef]
21. Przygrodzki, M.; Kubek, P. The Polish Practice of Probabilistic Approach in Power System Development Planning. *Energies* **2021**, *14*, 161. [CrossRef]
22. Choi, Y.; Song, D.; Yoon, S.; Koo, J. Comparison of Factorial and Latin Hypercube Sampling Designs for Meta-Models of Building Heating and Cooling Loads. *Energies* **2021**, *14*, 512. [CrossRef]
23. Wang, L.; Asomani, S.N.; Yuan, J.; Appiah, D. Geometrical Optimization of Pump-As-Turbine (PAT) Impellers for Enhancing Energy Efficiency with 1-D Theory. *Energies* **2020**, *13*, 4120. [CrossRef]
24. Agarwal, P.K.; Mustafa, N.H. K-Means Projective Clustering. In Proceedings of the Twenty-Third ACM SIGMOD-SIGACT-SIGART Symposium on Principles of Database Systems, Paris, France, 14–16 June 2004; Association for Computing Machinery: New York, NY, USA, 2004; pp. 155–165
25. Chalusiak, M.; Nawrot, W.; Buchaniec, S.; Brus, G. Swarm Intelligence-Based Methodology for Scanning Electron Microscope Image Segmentation of Solid Oxide Fuel Cell Anode. *Energies* **2021**, *14*, 3055. [CrossRef]
26. Tadjer, A.; Bratvold, R.B.; Hanea, R.G. Efficient Dimensionality Reduction Methods in Reservoir History Matching. *Energies* **2021**, *14*, 3137. [CrossRef]
27. Khan, A.N.; Iqbal, N.; Rizwan, A.; Ahmad, R.; Kim, D.H. An Ensemble Energy Consumption Forecasting Model Based on Spatial-Temporal Clustering Analysis in Residential Buildings. *Energies* **2021**, *14*, 3020. [CrossRef]
28. Lloyd, S. Least squares quantization in PCM. *IEEE Trans. Inf. Theory* **1982**, *28*, 129–137. [CrossRef]
29. Arthur, D.; Vassilvitskii, S. *k-means++: The Advantages of Careful Seeding*; Technical Report 2006-13; Stanford InfoLab: Stanford, CA, USA, 2006
30. Cheadle, C.; Vawter, M.P.; Freed, W.J.; Becker, K.G. Analysis of Microarray Data Using Z Score Transformation. *J. Mol. Diagn.* **2003**, *5*, 73–81. [CrossRef]
31. MathWorks. Cluster Visualization and Evaluation. Available online: https://se.mathworks.com/help/stats/cluster-evaluation.html?s_tid=CRUX_topnav (accessed on 1 June 2021).
32. Tibshirani, R.; Walther, G.; Hastie, T. Estimating the number of clusters in a data set via the gap statistic. *J. R. Stat. Soc. Ser. Stat. Methodol.* **2001**, *63*, 411–423. [CrossRef]

Article

Two-Phase Linear Hybrid Reluctance Actuator with Low Detent Force

Jordi Garcia-Amorós [1],*, Marc Marín-Genescà [2], Pere Andrada [3] and Eusebi Martínez-Piera [3]

[1]   Electrical and Electronic Engineering Departament, Universitat Rovira i Virgili, Av. Països Catalans 26,
       43007 Tarragona, Spain
[2]   Mechanical Engineering Department, Universitat Rovira i Virgili, Av. Països Catalans 26,
       43007 Tarragona, Spain; marc.marin@urv.cat
[3]   GAECE, Electric Engineering Department, Universitat Politècnica de Catalunya, BARCELONATECH,
       EPSEVG Av. Victor Balaguer 1, 08800 Vilanova i la Geltrú, Spain; pere.andrada@upc.edu (P.A.);
       mtzpiera@ee.upc.edu (E.M.-P.)
*   Correspondence: jordi.garcia-amoros@urv.cat

Received: 18 August 2020; Accepted: 1 October 2020; Published: 3 October 2020

Abstract: In this paper, a novel two-phase linear hybrid reluctance actuator with the double-sided segmented stator, made of laminated U cores, and an interior mover with permanent magnets is proposed. The permanent magnets are disposed of in a way that increases the thrust force of a double-sided linear switched reluctance actuator of the same size. To achieve this objective, each phase of the actuator is powered by a single H-bridge inverter. To reduce the detent force, the upper and the lower stator were shifted. Finite element analysis was used to demonstrate that the proposed actuator has a high force density with low detent force. In addition, a comparative study between the proposed linear hybrid reluctance actuator, linear switched reluctance, and linear permanent magnet actuators of the same size was performed. Finally, experimental tests carried out in a prototype confirmed the goals of the proposed actuator.

Keywords: linear electric actuators; linear switched reluctance actuators; permanent magnets; linear hybrid reluctance actuators; machine design; finite element analysis; detent force reduction

---

1. Introduction

Nowadays, many applications that use hydraulic or pneumatic drives or even electric drives combined with mechanical transmission systems are being substituted by linear electric actuators in industrial and aerospace applications. These actuators convert electric energy directly into a linear controlled movement with low cost and simple control. They are constituted of a fixed part or stator (primary) and a moving part or mover (secondary), both parts can contain coils, permanent magnets (PM) or bars. There are different kinds of linear electric actuators but linear switched reluctance motors (LSRM) are an attractive option due to their simple construction, robustness, and good fault capability despite their low force/mass ratio [1,2]. This disadvantage can be relieved by inserting permanent magnets (PM) in its magnetic structure, giving rise to the linear hybrid reluctance actuators. In some of these actuators, due to the disposition of the permanent magnets, the detent force is negligible [3]. Unfortunately, in most of these types of machines when the phases are not energized a detent force appears because of the interaction between the permanent magnets and the poles or teeth in which the windings are disposed of. Different techniques have been proposed for the reduction in detent force, the most usual are aside PM skewing and PM length/width adjustment [4,5], an asymmetric arrangement of PM, use of semi-closed slots [6], and utilization of auxiliary teeth or teeth notching [7]. Some other techniques use specific control strategies [8], in some cases, a combination of control and structural design (e.g., skewing PM, Halbach array) are employed [9,10], and others are based on the

---

shift of the permanent magnets, poles and slots, the slots or poles and the distance between the magnet segments of each pole [11–14].

During the work cycle (ascent or horizontal for linear machine), the propulsion force and direction of motion are in the same direction, and during descent, the propulsion force is acting in the direction opposite to that of the motion [15,16]. Propulsion force is a key concept for this type of machine, this defines the application in each type of machine configuration, and therefore defines a profile of force (F_X) and displacement (x) with peaks and minimums of propulsion force, which are necessary to adapt to the different applications [16]. Regarding the detent force importance, the main constraint of the slotted iron core type in magnet linear motors is the detent force that is caused by the interaction between the permanent magnet (PM) and the slotted iron core. This detent force will generate the pulsations of the propulsion force, and it will deteriorate the smoothness of motion drive, one of the main goals of the design of this type of electrical machine is to minimize this detent force. In general, the amplitude of the detent force is dependent on some major factors, such as pole-arc to a pole-pitch ratio of a magnet, air-gap length, slot opening length, skew of either stator teeth or magnet poles, and other key design factors [17,18].

Reluctance machines present good environmental behavior due to their high efficiency and inherent ease of assembly and dismantling [19]. For these reasons, among others, several studies have focused on new magnetic structures [20,21] in order to enhance their force performance [22] and increase force density by adding permanent magnets, some examples of electrical machines using permanent magnets have been developed recently in many research works [22–27]. As an example, linear switched reluctance motors (LSRM) and linear permanent magnet synchronous motors (LPMSMs) have been proposed for propelling a ropeless elevator [28], for an automotive suspension system [29], and for a linear generator in direct drive wave-power converter [30]. Despite their advantages, the permanent magnets linear (PML) motors exhibit some drawbacks: one of them is the presence of a cogging force, which can introduce a disturbance in positioning precision. The cogging force, in PML motors, is caused by two phenomena. The first one arises from the interaction between the PMs and the finite length of the armature core and is often called "end-effect". It can be minimized by adopting a suitable stator length [31] or modifying the extremity shape of the shorter part [32]. On the other hand, the U-core or U-channel air-core permanent magnet linear synchronous motors are widely applied in direct-drive linear motion servo systems, as they could offer significant advantages in terms of high efficiency, high positioning accuracy, rapid dynamic response, simple structure, and long service lifetime [33–36].

In this paper, a two-phase double-sided segmented stator, using laminated U cores, with an interior PM mover is proposed. Its main goal is to have high propulsion or thrust force and a low detente force.

The former concept-design of this two-phase linear hybrid reluctance motor comes from the optimized LSRM presented in [1], from which mover and stator dimensions are kept. That magnetic structure was hybridized by inserting Neodymium Iron Boron NdFeB magnets and was analyzed in [37].

The proposed two-phase actuator differs from the linear hybrid reluctance motor presented in [37] in the number of phases and in that the segmented stator is built with magnetically and mechanically decoupled U cores. This fact allows the relative displacement of the lower stator to respect the upper stator for reducing the detent force

After this introduction, the paper is organized as follows. In Section 2, a description of the proposed actuator is provided. The simulation of the actuator is performed with a 2D finite element analysis in Section 3. The reduction in the detent force is addressed in Section 4. A comparative study regarding switched reluctance and linear permanent actuators of the same size is performed in Section 5. The verification results obtained by 3D finite element analysis and experimental tests are shown in Section 6. Finally, in Section 7, conclusions from this research are drawn.

2. Description of the Proposed Actuator

In order to enhance the propulsion force of a conventional two-phase LSRM, a new two-phase double-sided linear hybrid reluctance actuator (PM-LHRM) is proposed. Figure 1 is a drawing of the PM-LHRM arrangement, wherein light grey shows the two symmetrical primary structures that contains the two phases, in blue and red, respectively, for the two phases B and C, whose magnetic circuit is made of two U laminated cores per phase. The secondary, depicted in grey, has disposed of the permanent magnets (NdFeB-N32) (in black/white) between their poles with the direction of magnetization shown in Figure 1 (*S-N-S-N-S . . .* in both sides). Each phase is energized through an H-bridge inverter that allows switching the phase current to a positive value +I to a negative value –I in the period of conduction of the corresponding phase. The goal is to double the total average force of a linear switched reluctance actuator of the same geometry. The actuator is intended for short-time duty cycle (S2) applications with an average total thrust of 100 N at a speed of 0.15 m/s, which not require a very precise positioning such as automatic door opener systems.

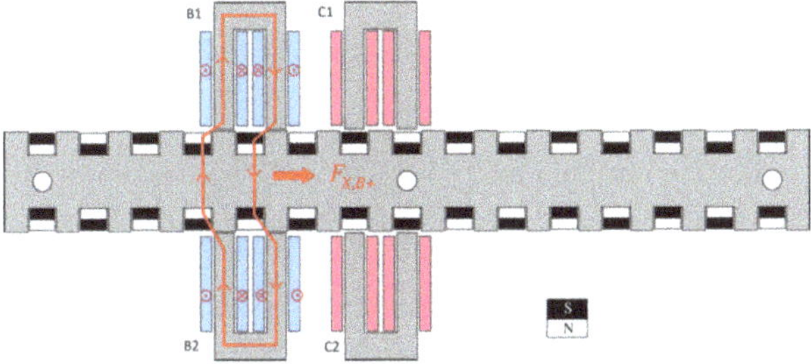

Figure 1. View of the proposed two-phase double-sided linear hybrid reluctance actuator (PM-LHRM) and operation principle when exciting phase B+ at $x = 0$ mm.

The operating principle is shown in Figure 1, which sketches a field line when exciting phase B with a positive current, called B+, and the resulting propulsion force ($F_{X,B+}$) in the right direction (positive). Without PMs in the mover, the structure is a conventional LSRM, in which feeding the same phase B+ the propulsion force is zero at $x = 0$ mm, and the same would happen by feeding phase C. The existence of these zero-force positions disables the two-phase LSRM as a propulsion actuator.

In order to assess the PM-LHRM, it will be compared with a pure reluctance motor, demoted by LSRM (see Figure 2a), and with a permanent magnet linear motor without iron poles in the mover, denoted by PM-LM (see Figure 2b). The LSRM and PM-LM actuators have the same stator structure of Figure 1, but their respective movers are different as shown in Figure 2.

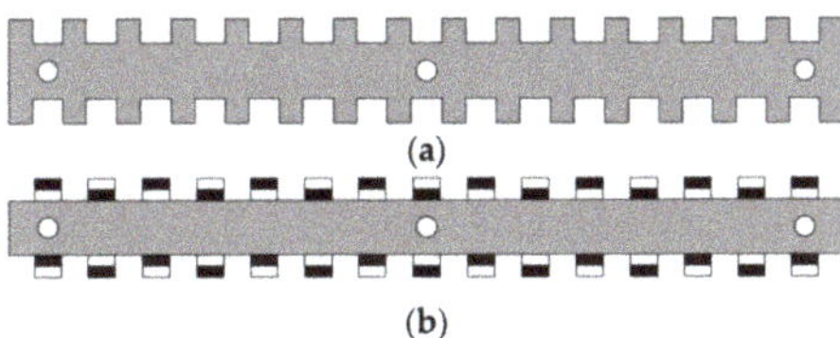

Figure 2. Mover structures (**a**) linear switched reluctance motor (LSRM) (**b**) PM-LM.

Some relevant details about the design of the proposed machine are given in the Appendix A.

3. Simulation of the Actuator

The PM-LHRM, LSRM, and PM-LM actuators were simulated using 2D finite element analysis [38] and *Matlab*, for a current density of 10 A/mm^2. This current density value is chosen because the actuator is conceived for short intermittent duty cycles. It is important to point out that there is only one flux-path per phase, and the phases are magnetically uncoupled, that is, the flux created by coil phase B does not link coil phase C. The simulation model has more than 185k elements and nearly 93k nodes, and a simulation solver precision of 10 nano. Figure 3a shows the comparison between the force profiles of the three structures, in which it can be seen a prominent force peak (>150 N) for the PM-LHRM, which doubles the force peak (F_{peak}) of the LSRM. Moreover, the force becomes anti-symmetric for the PM-LHRM when the phase-current is inverted, that means zero-force positions can be overlapped when phases are fed with the appropriate sequence (e.g., B+ C+ B- C-), as can be seen in Figure 3b. Therefore, adding the PMs to the two-phase LSRM magnetic circuit allows the elimination of zero-force positions, which position this PM-LHRM configuration as a good solution for high-density thrust actuators. The relevant values shown in Figure 3 are F_{peak} = 162.8 N for the PM-LHRM, F_{peak} = 128.8 N for PM-LM, F_{peak} = 75.8 N for LSRM, and the average force values F_{ave} = 91.4 N, F_{ave} = 78.1 N, F_{ave} = 46.6 N, respectively, being the average force values computed over a semi-period.

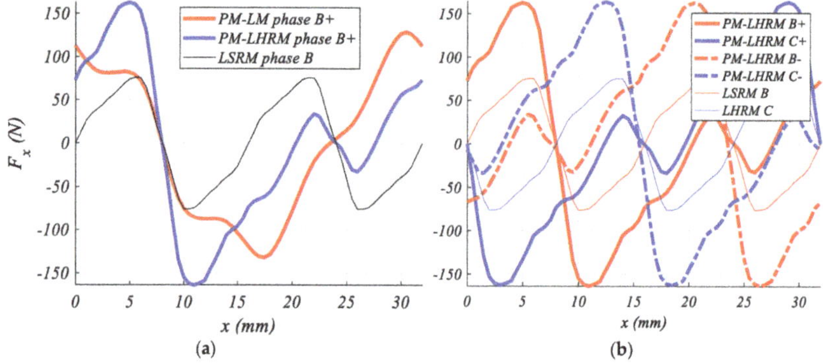

Figure 3. Propulsion force (N) vs. position (mm) simulation results at J = 10 A/mm^2 and shift = 0 mm. (**a**) Comparison of the 3 structures. (**b**) Comparison for the PM-LHRM and LSRM structure feeding phases B+, C+, B-, C-.

The detent force that appears when currents are zero, due to the presence of PMs, is represented in Figure 4. The detent force values are 28.2 N of peak, (17.3% of peak force) and a root mean square (*rms*) value 18.9 N-rms for the PM-LHRM structure, and 22.9 N of peak (17.8% of peak force) and 14.6 N-rms for PM-LM. These values are far from negligible and therefore they are a significant drawback, despite the better performance of PM-LHRM and PM-LM in respect to the LSRM.

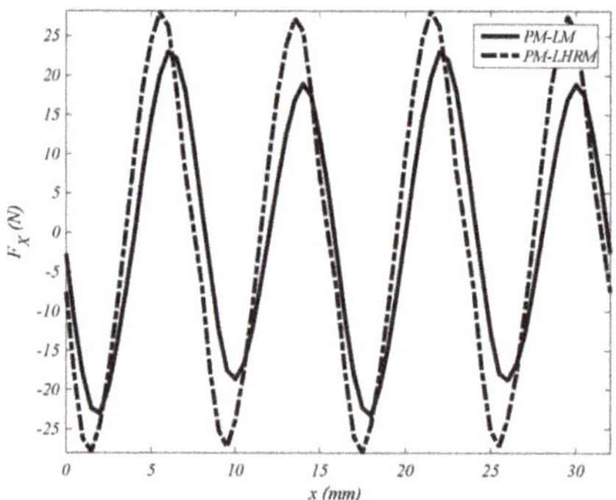

Figure 4. Detent force or cogging force for PM-LM and PM-LHRM at shift = 0 mm.

4. Reduction in Detent Force

The PM linear motors and by extension the hybrid magnetic structures exhibit the so-called detent force, which means the existence of a threshold force under which the motor cannot operate. This force is due to the PM flux lines that sew stator and rotor at given positions due to the slotted structures, called "slotting effect". This causes error in positioning, vibrations, and noise. Reference [12] classifies linear motors according to X configuration, where the secondary is longer than primary, and Y configuration, where secondary is shorter than primary. Both configurations suffer the "slotting effect", additionally X configuration also has the "end-effect", which has less relevance than the "slotting effect", as shown in Section 6.

After considering the different alternatives exposed in Section 1 for the reduction in detent force, the option based on the technique of pole shifting [12] was selected because it was the easiest to implement given the modular arrangement of the proposed actuator built with independent U cores. This was implemented maintaining the U cores of the upper stator—B1 and C1—in the same position (see Figure 1), and displacing the U cores of the lower stator—B2 and C2—a certain distance, called shift, into the right direction.

To analytically approach the analysis, several methodologies are presented in [9,10,14]. In this case, a simplified magnetic equivalent circuit based on Figure 1 is proposed (see Figure 5). In this approach, the iron reluctances are neglected as well as the leakage fluxes. In Figure 5, FI stands for the phase-coil magneto-motive force (*mmf*) and FM for the magnet's *mmf*, being $F_M = H_M \cdot l_m$. A sinusoidal flux variation is also assumed in respect to position x, see Equations (1) and (2) for phase B and Equations (3) and (4) for phase C. It is also considered a variable accounting for the displacement s between the upper and lower phase-stators to reduce the detent force. The numerical values for the variables T_s, L_W and C_m are given in the Appendix A Table A1.

$$\phi_{B1}(x) = B_r \cdot L_W \cdot C_m \cdot sin\left(\frac{\pi \cdot x}{T_S}\right) \tag{1}$$

$$\phi_{B2}(x,s) = B_r \cdot L_W \cdot C_m \cdot sin\left(\frac{\pi \cdot (x+s)}{T_S}\right) \tag{2}$$

$$\phi_{C1}(x) = B_r \cdot L_W \cdot C_m \cdot cos\left(\frac{\pi \cdot \left(x - \frac{T_S}{2}\right)}{T_S}\right) \tag{3}$$

$$\phi_{C2}(x,s) = B_r \cdot L_W \cdot C_m \cdot cos\left(\frac{\pi \cdot \left(x - \frac{T_S}{2} + s\right)}{T_S}\right) \tag{4}$$

The reluctances, $\mathcal{R}(x,s)$, of each U core, with regards to the value of x and s are obtained in the following Equations (5)–(8):

$$\mathcal{R}_{B1}(x,s) = \left(\frac{\mathcal{R}_{max} - \mathcal{R}_{min}}{2}\right) \cdot \left(1 - cos\left(\frac{2 \cdot \pi \cdot x}{T_S}\right)\right) + \mathcal{R}_{min} \tag{5}$$

$$\mathcal{R}_{B2}(x,s) = \left(\frac{\mathcal{R}_{max} - \mathcal{R}_{min}}{2}\right) \cdot \left(1 - cos\left(\frac{2 \cdot \pi \cdot (x+s)}{T_S}\right)\right) + \mathcal{R}_{min} \tag{6}$$

$$\mathcal{R}_{C1}(x,s) = \left(\frac{\mathcal{R}_{max} - \mathcal{R}_{min}}{2}\right) \cdot \left(1 - cos\left(\frac{2 \cdot \pi \cdot \left(x - \frac{T_S}{2}\right)}{T_S}\right)\right) + \mathcal{R}_{min} \tag{7}$$

$$\mathcal{R}_{C2}(x,s) = \left(\frac{\mathcal{R}_{max} - \mathcal{R}_{min}}{2}\right) \cdot \left(1 - cos\left(\frac{2 \cdot \pi \cdot \left(x - \frac{T_S}{2} + s\right)}{T_S}\right)\right) + \mathcal{R}_{min} \tag{8}$$

The reluctance value $\mathcal{R}_{max}$ corresponds to the unaligned position (see phase C in Figure 1) and $\mathcal{R}_{min}$ to the aligned position (see phase B in Figure 1), these two reluctance values can be obtained analytically by computing and adding the permeance flux-tubes set obtained at the given positions. The detent force (F_d) is then obtained by Equation (9).

$$F_d(x,\ s) = \frac{1}{2}\cdot\sum_{i=1}^{2}\phi_{Bi}(x,s)^2\cdot\frac{\delta\mathcal{R}_{Bi}(x,\ s)}{\delta x} + \frac{1}{2}\cdot\sum_{i=1}^{2}\phi_{Ci}(x,s)^2\cdot\frac{\delta\mathcal{R}_{Ci}(x,\ s)}{\delta x} \tag{9}$$

Thus, F_d is minimized regarding the shift variable, s, reaching a minimum value of detent force at $s = 4$ mm for the given values.

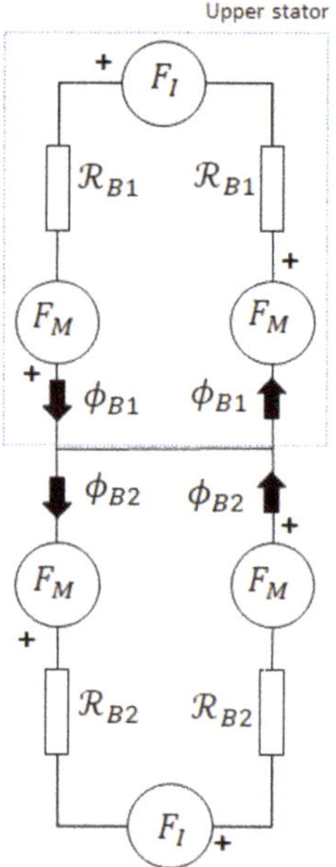

Figure 5. Simplified phase-B magnetic equivalent circuit for PM-LHRM.

In order to provide an expression involving the phase-current (I) and the number of turns (N), it can be obtained by applying the second *Kirchoff* law to either of the loops (see Figure 5), resulting:

$$F_I = 2\cdot R_{B1}\cdot\phi_{B1} + 2\cdot F_M \rightarrow I = \frac{2}{N}\cdot(R_{B1}\cdot\phi_{B1} + H_M\cdot l_M) \tag{10}$$

The detent force is computed numerically using 2DFEM solver [39] by using Maxwell stress tensor and analytically by the set of Equations (1)–(9). Figure 6a shows the flux line distribution for a given position considering only phase B. Figure 6b shows the flux distribution of the whole PM-LHRM for a given shift. To assess the goodness of the analytical approach, 2DFEM simulations are computed over a one phase (B), and it is considered the detent force of other phase (C) out of phase (i.e., 180°), regarding to phase B. The detent force comparison of each phase (phase B in red and phase C in blue) and the total force (in black) is depicted in Figure 7, for both analytical and 2DFEM results. As can be seen, both models match quite well at $s = 4$ mm when both, FEM and analytical, predict a minimum detent force. The differences with other shift values are due to the analytical model's accuracy, few flux tubes are considered, since the goal is not to present an accurate analytical model but rather to give an analytical vision on the effects of the shift displacement on the detent force.

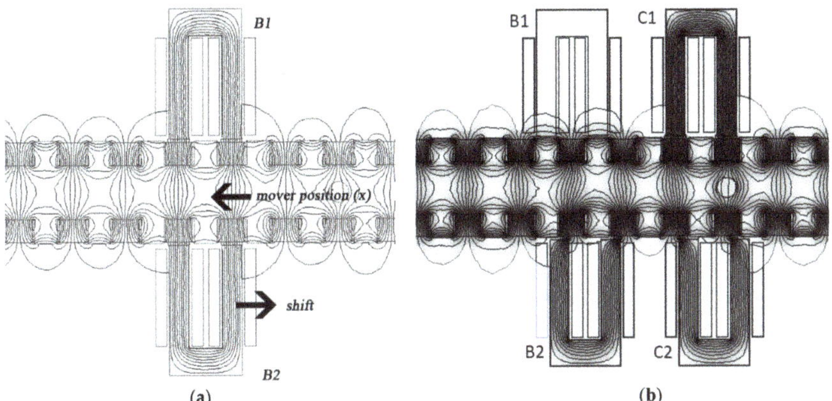

Figure 6. Field line distribution for PM-LHRM, $I_B = I_C = 0$ and position $x = 0$ mm. (**a**) B-phase $s = 0$ mm. (**b**) whole PM-LHRM $s = 4$ mm.

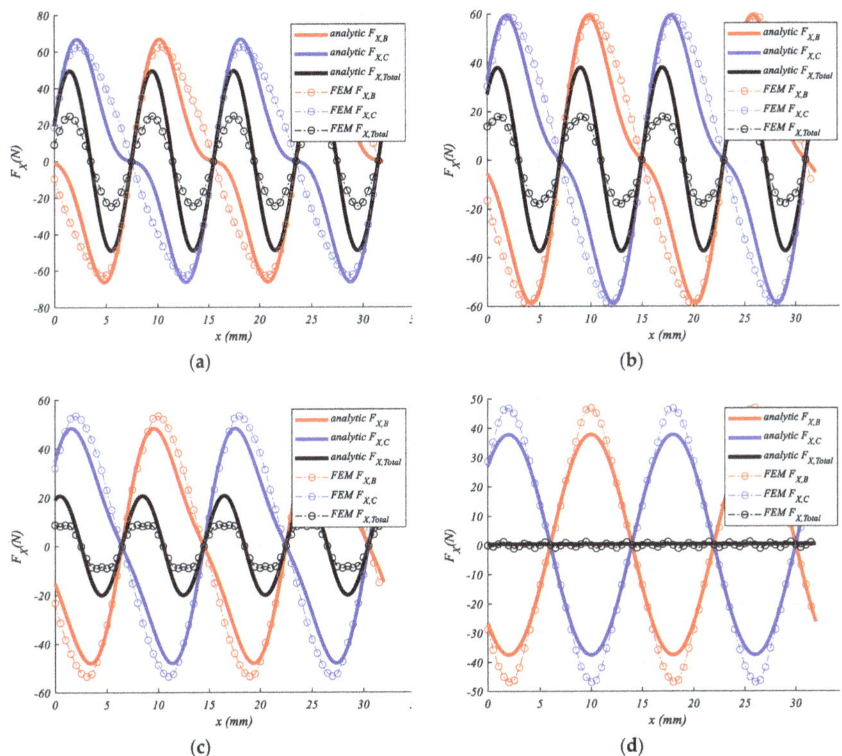

Figure 7. Detent force comparison results. (**a**) $s = 1$ mm. (**b**) $s = 2$ mm. (**c**) $s = 3$ mm. (**d**) $s = 4$ mm.

Figure 8 shows the simulation results of the detent force versus the shift for the whole PM-LHRM. The detent force results are minimal $s = 4$ mm as predicted by the analytical model, see Figure 7d. The detent force is computed for the whole PM-LHRM for ranges of shift (s) and mover positions (x), which are: $s \in [0 \div 6$ mm, $\Delta s = 1$ mm] and $x \in [0 \div 32$ mm, $\Delta x = 0.5$ mm]. The mover runs to the left, being the initial position, $x = 0$, the position shown in Figure 1. The distance covered by the mover, from $x = 0$ mm (see Figure 1) to the position in which stator U core B1 reaches the same relative position of C1 at $x = 0$ mm, is the period of detent force (τ), which for this actuator is $\tau = 8$ mm. The results corroborate the analytical results obtained from (9), and the minimum detent force is obtained for a shift of $s = 4$ mm (see Figure 7). Repeating the simulations for the PM-LM structure also finds a minimum in the detent force at the same shift of 4 mm.

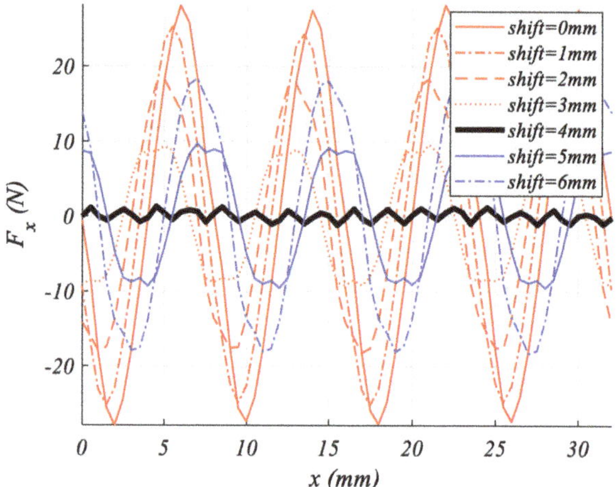

Figure 8. Detent force simulation results of the whole PM-LHRM.

5. Comparative Study for Shifted Structures

A comparison between the proposed PM-LHRM and the PM-LM simulation results is presented. Both motors are of the same size and have the same shift of $s = 4$ mm. Figure 9a shows the detent force comparison of the two structures, in which the PM-LHRM present a significant lower detent force, meanwhile PM-LM has a wide cycle of the force of 16 mm, with peaks points over 2.5 N and −3 N. Figure 9b shows the propulsion force comparison results of the PM-LM and PM-LHRM structures, and also the LSRM. As can be seen, the force distributions obtained after shifting are significantly different in both PM motors. The peak forces are 130.4 N for PM-LHRM and 105.7 N for PM-LM, and the average forces are 60 N and 69 N, respectively. From the simulation results presented in Figure 9a,b, PM-LHRM exhibits a lower detent force and a higher force peak in comparison with PM-LM. The LSRM shows the lowest propulsion force peak at 75 N.

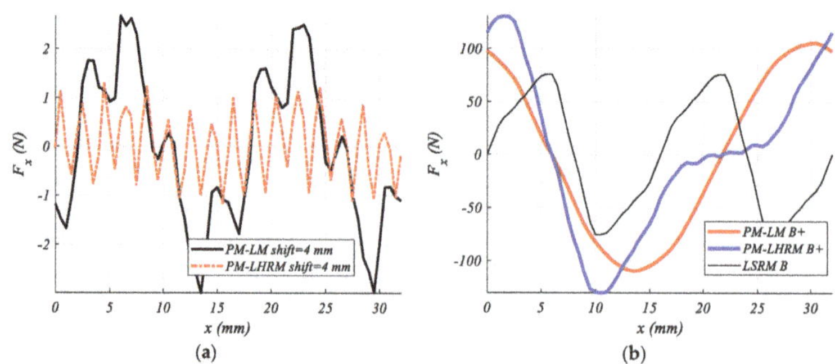

Figure 9. (**a**) Detent force comparison results at $s = 4$ mm. (**b**) Phase force comparison results for a current density of 10 A/mm^2 and shift = 4 mm.

In order to assess the goodness of the proposed actuator versus its PM-LM counterpart, Figure 10a,b compare their propulsion force profiles, obtained by feeding the phases with a flat current waveform in switched reluctance mode, that is without overlapping phase currents. The phases are fed for obtaining a positive trust in the sequence B+ C- B- C+. The total trust obtained is the enveloping of the force profiles in the red line.

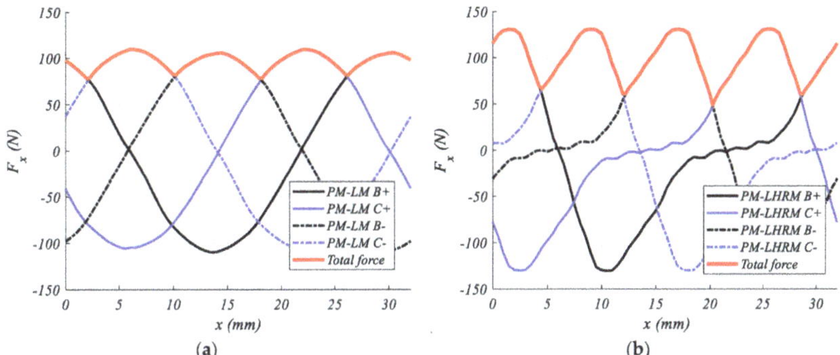

Figure 10. Phase force profiles and total force. (**a**) PM-LM at shift = 4 mm. (**b**) PM-LHRM at shift = 4 mm.

Figure 11 shows the comparison results of the total propulsion force of the three motors analyzed. The main values are summarized in Table 1, where to complete the study ripple factor is included (see Equation (11)), having been obtained through the following expression:

$$F_{ripple} = \frac{F_{x,max} - F_{x,min}}{F_{x,ave}} \tag{11}$$

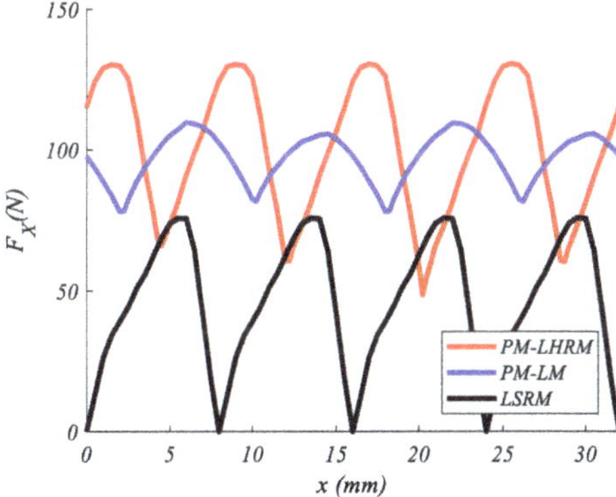

Figure 11. Total force simulation comparison results for the PM-LHRM and PM-LM, s = 4 mm, and the LSRM at 10 A/mm^2.

Table 1. Main comparison simulation results from PM-LHRM, PM-LM, and LSRM.

Structure	Total Force			Phase Force		Detent Force	
	F_{peak}	F_{ave}	F_{ripple}	F_{peak}	F_{ave}	F_{peak}	F_{rms}
PM-LHRM	130.7 N	100.8 N	0.81	130.7 N	60 N	1.3 N	0.67 N
PM-LM	109.8 N	97.1 N	0.33	105.7 N	69.7 N	2.66 N	1.54 N
LSRM	75.8 N	46.2 N	1.64	75.8 N	46.5 N	0	0

In Figure 11, the work cycle of propulsion force shows a maximum for PM-LHRM, 130 N. For PM-LM shows a propulsion force with a displacement cycle of 8 mm, with a lower maximum of force than the LHRM configuration (109.8 N), and minimum at 80 N. Finally, LSRM shows the lowest propulsion force than the three configurations (75.8 N), and a displacement cycle of 8 mm.

Table 1 collects the three parameters (peak force, average force, and ripple force) obtained from Figure 11, as well as the detent force peak and rms values obtained from Figure 9a. PM-LHRM shows the highest force peak, on the other hand, LSRM presents the lowest peak force, similarly, the F_{ave}, presents the same trend as the F_{peak} profiles. Detent forces present higher values for PM-LM (2.66 N) than PM-LHRM (1.3 N), but in very low percent values from F_{peak} analyzed: 2.4% to 1%, for PM-LM and PM-LHRM configurations, respectively.

6. Verification Results

A prototype of the proposed PM-LHRM actuator was built. In Figure 12a, a lamination of stator U-cores and the mover are shown. A picture of one of the U-core wounded is depicted in Figure 12b. The complete actuator can be seen in Figure 12c. The design specifications of the linear hybrid reluctance actuator are those given in the appendix.

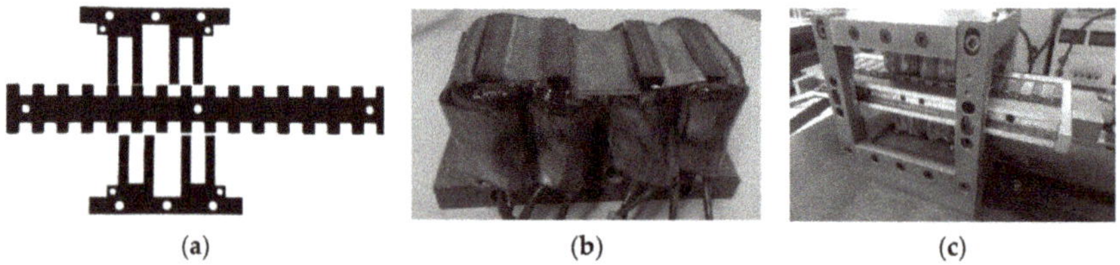

(a)   (b)   (c)

Figure 12. (a) Lamination of U-cores and mover. (b) Detail of one U-core with the stator coils. (c) View of the complete actuator.

The force measurements were done with a load cell, UTILCELL CR200. On the other hand, the magnetic field of all the permanent magnets were measured on the airgap surface with a gaussmeter model Tenmars TM-197 AC/DC, giving values ranging from ± 0.47 to ± 0.51, giving an acceptable magnetic field symmetry.

To validate the FEM simulations, a 3D finite element analysis using Altair Flux™ [40] was performed to account for end-effects. An experimental test set-up (see Figure 12c) was used to obtain the curve force profile for each phase of the prototype (see Figures 13 and 14), feeding one current at a time: I_{B+} I_{B-} I_{C+} I_{C-}, which for J = 10 A/mm^2 results from a value of 3.12 A. These results are presented in Figures 13 and 14 along with the 2DFEM simulation results for each phase for a shift displacement of 0 and 4 mm, respectively. The test system positioning consisted of a calibrated screw (1 mm per revolution) with a range of 28 mm. From these figures, it can be seen a good agreement between simulation and experimental results. It is appreciated there is a slight shift of the experimental values in respect to the simulation, this is due to a positioning error of the screw system.

Figure 15 shows comparison results of the detent force for the two-shift considered. As can be seen, the experimental values of detent force for s = 4 mm reveals a significant difference between the 2DFEM results.

To shed some light on this discrepancy, a 3DFEM [41] simulation was performed. For s = 0 mm, there is an acceptable concordance between the results 2DFEM, 3DFEM, and experimental (see Figure 16a). In contrast, for s = 4 mm the differences persist (see Figure 16a), which leads to the belief that end-effects are not relevant enough to produce such difference.

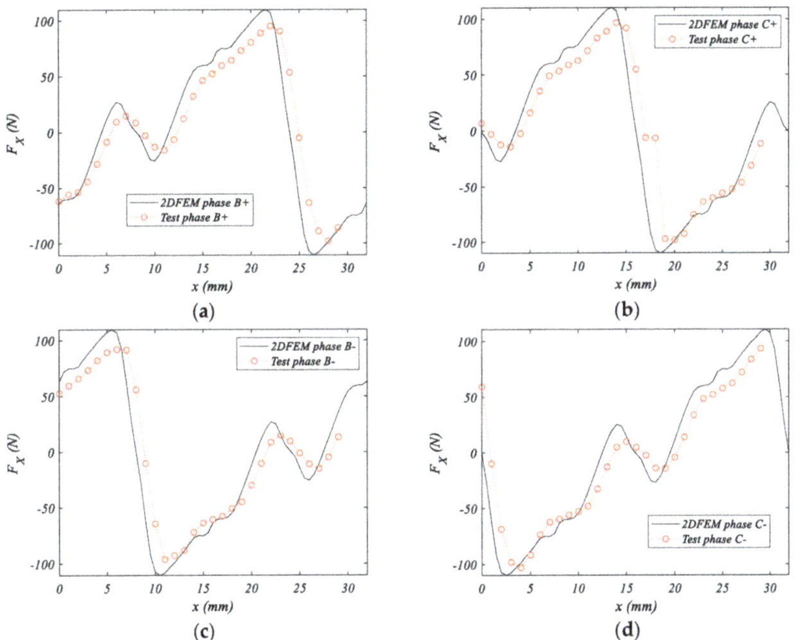

Figure 13. Propulsion force comparison results for J = 10 A/mm², shift = 0 mm (**a**). Phase I_B = 3.12A. (**b**) Phase I_C = 3.12A. (**c**) Phase I_B = −3.12A. (**d**) Phase I_C = −3.12A.

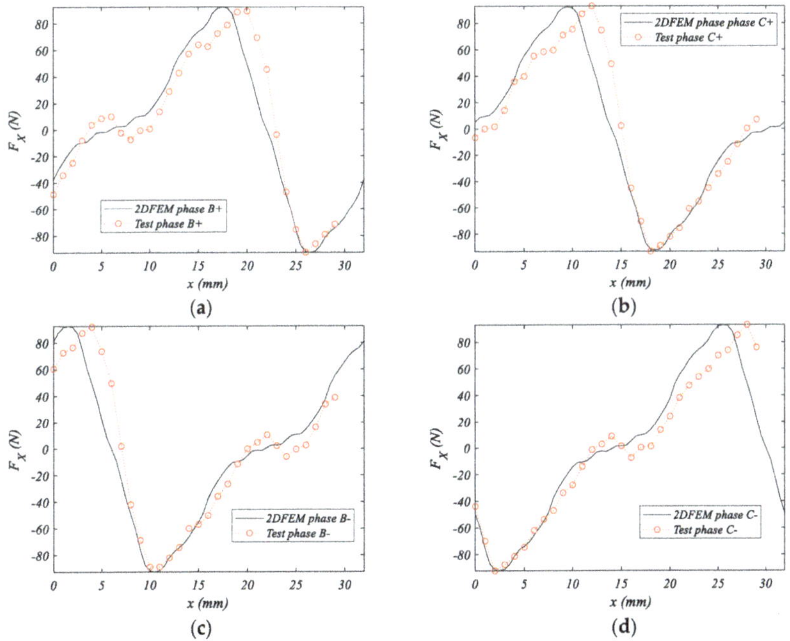

Figure 14. Propulsion force comparison results for J = 10 A/mm², shift = 4 mm (**a**). Phase I_B = 3.12A. (**b**) Phase I_C = 3.12A. (**c**) Phase I_B = −3.12A. (**d**) Phase I_C = −3.12A.

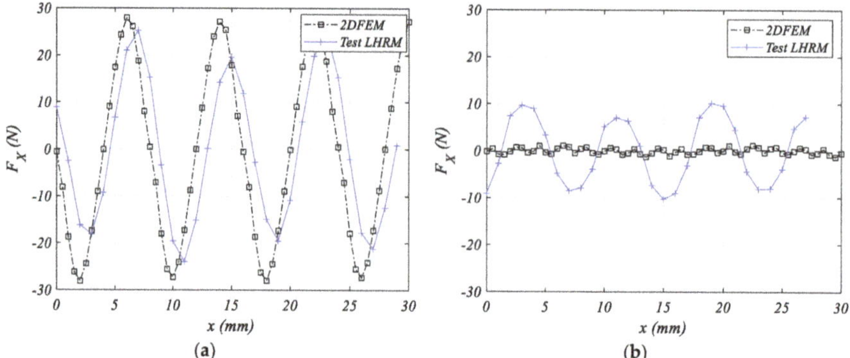

Figure 15. 2DFEM vs. experimental detent force comparison results. (**a**) $s = 0$ mm. (**b**) $s = 4$ mm.

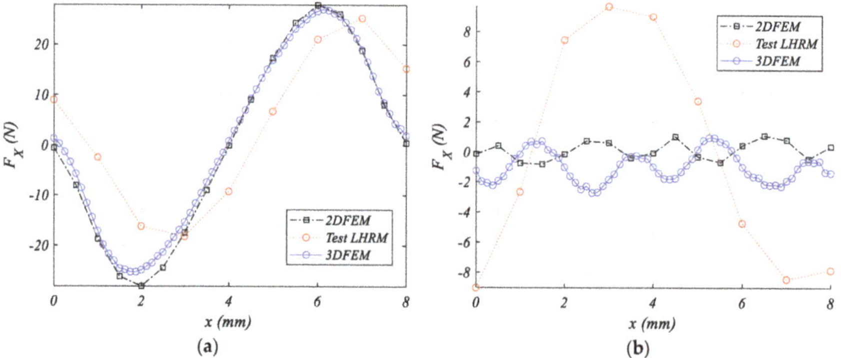

Figure 16. Detent force comparison results. (**a**) $s = 0$ mm. (**b**) $s = 4$ mm.

Such deviation in detent force for $s = 4$ mm can be due to the combination of three main causes. The first one is due to static friction, which is not accounted for in FEM simulations. The second cause is due to the mechanical assembly tolerances of the prototype. This affects two crucial aspects: (a) different airgap lengths due to constructive defects, and (b) the lower stator shift tolerance. A FEM sensitivity analysis has been carried out for the airgap tolerances (lengths), that is, the upper airgap $g_1 = 0.5 \pm \Delta g$, the lower airgap $g_2 = 0.5 \pm \Delta g$, and by taking a tolerance range of $\Delta g \in [-0.4, 0.4]$ mm. The curve patterns of detent force obtained from sensitivity analysis fit with the experimental measurement detent force pattern when $g_1 \neq g_2$ (see Figure 15b blue line), which indicates the probable existence of this kind of asymmetry, since only for $g_1 = g_2$ the detent force is minimum ($s = 4$ mm) and follows the pattern shown in Figure 8 (black line). The influence of the tolerance of the shift of lower semi stator can be seen in Figure 8. For instance, for $s = 3$ mm (see Figure 8) the detent force pattern matches the experimental detent force (see Figure 15b). These two observations, different airgap lengths and shift tolerances, lead to the conclusion of the existence of some mechanical asymmetry, not easy to quantify, which affects these highly sensitive parameters (g_1, g_2, s) and produces this significant difference between FEM (2D and 3D) and experimental measures. The third cause of perturbation can be assigned to the end-effects phenomena, which produces a magnetic force at the stator edges [38,41].

Despite this deviation in the experimental and FEM results, it is important to point out the relevant reduction (>50%) in the detent force when one stator is shifted from $s = 0$ to $s = 4$ mm (see Figure 15), which validates the methodology and the PM-LHRM as an interesting choice in high-density force applications.

7. Conclusions

In this paper, a novel type of two-phase hybrid reluctance actuator (PM-LHRM) is presented. The stator is double-sided and consists of four laminated U cores, two of them are placed on the upper side and the two others on the lower side. The permanent magnets are placed into the mover between their poles and have been disposed of with a determined magnetization. Each phase is energized by its H-bridge to allow the switching of the phase current to a positive value (+I) and to a negative value (–I) in the period of conduction of each phase. The U cores of the lower stator are shifted in respect to the U cores of the upper stator for reducing the detent force. Simulations demonstrate that this actuator doubles the total thrust of an LSRM actuator of the same size, with a reduced detent force. The PM-LHRM was compared with a PM-LM (no poles in the mover), resulting in a lower detent force for $s = 4$ mm in the PM-LHRM structure and a higher peak and average force, 19% and 3.8%, respectively.

From the results obtained, it can be drawn that the PM-LHRM structure has a better performance since applying a convenient force control (i.e., feeding with an appropriate current waveform), the force the ripple factor can be minimized, and the average propulsion force fit to its maximum value. It has also been revealed that the detent force of this actuator is highly sensitive to constructive defects, especially the airgaps length and the stator shift displacement.

Author Contributions: Conceptualization, J.G.-A.; methodology, J.G.-A. and P.A.; software, J.G.-A. and E.M.-P.; validation, J.G.-A., E.M.-P. and M.M.-G.; formal analysis, J.G.-A. and P.A.; investigation, J.G.-A.; resources, J.G.-A. and P.A.; data curation, M.M.-G.; writing—original draft preparation, M.M.-G.; writing—review and editing, J.G.-A. and P.A.; supervision, J.G.-A. All authors have read and agreed to the published version of the manuscript.

Funding: This research received no external funding.

Acknowledgments: To the collaboration of the company ARSAN-ESTAMPACIONES (Huarte-Pamplona, Spain).

Conflicts of Interest: The authors declare no conflict of interest.

Nomenclature

b_p	Stator pole width (m)
b_s	Mover pole width (m)
c_m	Permanent magnet width (m)
c_p	Stator slot width (m)
c_s	Mover slot width (m)
F_x	Propulsion force or thrust (N)
F_d	Detent force (N)
g	Air gap length (m)
H_c	Permanent magnet coercivity (kA/m)
h_y	Stator yoke height (m)
l_m	Permanent magnet length (m)
l_p	Stator pole length (m)
l_s	Mover pole length (m)
m	Number of phases
N_{ph}	Number of coils per phase
ψ	Flux linkage (Wb)
s	Shift (m)
τ	Period of detent torque (m)
x	Mover position (m)

Appendix A

The number of wires per pole are 38, the wire diameter is 1.4 mm, the number of coils per phase are four, and the material type of linear machine is specific magnetic steel type: M 270-50A, finally, the permanent magnets (PM) used in the linear switched machine are of neodymium material type, with specific denomination NdFeB-32.

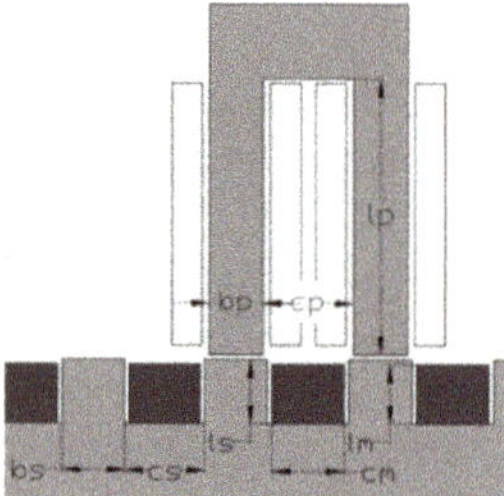

Figure A1. PM-LHRM main dimensions.

Table A1. PM-LHRM main dimension values.

Mover pole width	b_s	7 mm
Mover slot width	c_s	9 mm
Mover slot height	l_s	7 mm
Permanent magnet length	l_m	6.5 mm
Permanent magnet width	c_m	8.5 mm
Stator pole width	b_p	6 mm
Stator slot width	c_p	10 mm
Stator pole length	l_p	30 mm
Number of wires per pole	N	38
Wire diameter	d	1.4 mm
Number of coils per phase	N_{ph}	4
Magnetic steel	M 270-50A	
Permanent magnet	NdFeB-32	

References

1. Amoros, J.G.; Andrada, P. Sensitivity Analysis of Geometrical Parameters on a Double-Sided Linear Switched Reluctance Motor. *IEEE Trans. Ind. Electron.* **2010**, *57*, 311–319.
2. García, A.J.; Andrada, P.; Blanque, B.; Marin-Genesca, M. Influence of Design Parameters in the Optimization of Linear Switched Reluctance Motor under Thermal Constraints. *IEEE Trans. Ind. Electron.* **2018**, *65*, 1875–1883.
3. Szabó, L.; Viorel, I.A. On a high force modular surface motor. In Proceedings of the 10th International Power Electronics and Motion Control Conference (EPE-PEMC), Dubrovnik, Croatia, 9–11 September 2002.
4. Lim, K.C.; Woo, J.K.; Kang, G.H.; Hong, J.P.; Kim, G.T. Detent Force Minimization Techniques in Permanent Magnet Linear Synchronous Motors. *IEEE Trans. Magn.* **2002**, *38*, 1157–1160. [CrossRef]
5. Jang, S.; Lee, S.; Yoon, I. Design Criteria for Detent Force Reduction of Permanent-Magnet Linear Synchronous Motors with Halbach Array. *IEEE Trans. Magn.* **2002**, *38*, 3261–3263. [CrossRef]
6. Sung, W.S.; Gang, H.J.; Min, M.K.; Jang, Y.C. Characteristic Analysis of the Influence of Auxiliary Teeth and Notching on the Reduction of the Detent Force of a Permanent Magnet Linear Synchronous Machine. *IEEE Trans. Appl. Supercond.* **2018**, *28*, 1–5. [CrossRef]
7. Bascetta, L.; Rocco, P.; Magnani, G. Force Ripple Compensation in Linear Motors Based on Closed-Loop Position-Dependent Identification. *IEEE/ASME Trans. Mechatron.* **2010**, *15*, 3. [CrossRef]
8. Yu-Wu, Z.; Yun-Huyn, C. Thrust Ripples Suppression of Permanent Magnet Linear Synchronous Motor. *IEEE Trans. Magn.* **2007**, *43*, 2537–2539. [CrossRef]
9. Nevaranta, N.; Huikuri, M.; Niemelä, M.; Pyrhönen, J. Cogging force compensation of a discontinuous permanent magnet track linear motor drive. In Proceedings of the European Conference on Power Electronics and Applications (EPE'17 ECCE Europe), Warsaw, Poland, 11–14 September 2017. [CrossRef]

10. Wang, Q.; Zhao, B.; Zou, J.; Li, Y. Minimization of Cogging Force in Fractional-Slot Permanent Magnet Linear Motors with Double-Layer Concentrated Windings. *Energies* **2016**, *9*, 918. [CrossRef]
11. Wang, M.; Li, L.; Pan, D. Detent Force Compensation for PMLSM Systems Based on Structural Design and Control Method Combination. *IEEE Trans. Ind. Electron.* **2015**, *62*, 11. [CrossRef]
12. Bianchi, N.; Bolognani, S.; Cappello, A.D.F. Reduction of cogging force in PM linear motors by pole-shifting. *IEEE Proc. Electr. Power Appl.* **2005**, *152*, 703–709. [CrossRef]
13. Kwon, Y.; Kim, W. Steady-State Modeling and Analysis of a Double-Sided Interior Permanent-Magnet Flat Linear Brushless Motor with Slot-Phase Shift and Alternate Teeth Windings. *IEEE Trans. Magn.* **2016**, *52*, 11. [CrossRef]
14. Setiabudy, R.; Herlina; Putra, Y.S. Reduction of cogging torque on brushless direct current motor with segmentation of magnet permanent. In Proceedings of the International Conference on Information Technology, Computer, and Electrical Engineering (ICITACEE), Semarang, Indonesia, 18–19 October 2017; pp. 81–86. [CrossRef]
15. Lim, H.S.; Krishnan, R.; Lobo, N.S. Design and control of a linear propulsion system for an elevator using linear switched reluctance motors. *IEEE Trans. Ind. Electron.* **2005**, *55*, 1584–1591.
16. Bae, H.-K.; Lee, B.-S.; Vijayraghavan, P.; Krishnan, R. A linear switched reluctance motor: Converter and control. *IEEE Trans. Ind. Appl.* **2000**, *36*, 1351–1359.
17. Lee, S.; Kim, S.; Saha, S.; Zhu, Y.; Cho, Y. Optimal Structure Design for Minimizing Detent Force of PMLSM for a Ropeless Elevator. *IEEE Trans. Magn.* **2014**, *50*, 1–4. [CrossRef]
18. Jahns, T.M.; Soong, W.L. Pulsating torque minimization techniques for permanent magnet AC motor drives—A review. *IEEE Trans. Ind. Electron.* **1996**, *43*, 321–330. [CrossRef]
19. Andrada, P.; Blanque, B.; Martinez, E.; Perat, J.I.; Sanchez, J.A.; Torrent, M. Environmental and life cycle cost analysis of one switched reluctance motor drive and two inverter-fed induction motor drives. *IET Electr. Power Appl.* **2012**, *6*, 390–398. [CrossRef]
20. Chen, H.; Nie, R.; Yan, W.A. Novel Structure Single-Phase Tubular Switched Reluctance Linear Motor. *IEEE Trans. Magn.* **2017**, *53*, 1–4. [CrossRef]
21. Zhao, W.; Zheng, J.; Wang, J.; Liu, G.; Zhao, J.; Fang, Z. Design and Analysis of a Linear Permanent- Magnet Vernier Machine with Improved Force Density. *IEEE Trans. Ind. Electr.* **2016**, *63*, 2072–2082. [CrossRef]
22. Enrici, P.; Dumas, F.; Ziegler, N.; Matt, D. Design of a High-Performance Multi-Air Gap Linear Actuator for Aeronautical Applications. *IEEE Trans. Energy Convers.* **2016**, *31*, 896–905. [CrossRef]
23. Pan, J.F.; Wang, W.; Zhang, B.; Cheng, E.; Yuan, J.; Qiu, L.; Wu, X. Complimentary Force Allocation Control for a Dual-Mover Linear Switched Reluctance Machine. *Energies* **2018**, *11*, 23. [CrossRef]
24. Andrada, P.; Blanqué, B.; Martínez, E.; Torrent, M.; Garcia-Amorós, J.; Perat, J.I. New Linear Hybrid Reluctance Actuator. In Proceedings of the International Conference on Electrical Machines (ICEM), Berlin, Germany, 1–4 September 2014.
25. Andrada, P.; Blanque, B.; Martinez, E.; Torrent, M. A Novel Type of Hybrid Reluctance Motor Drive. *IEEE Trans. Ind. Electr.* **2014**, *61*, 4337–4345. [CrossRef]
26. Ullah, S.; McDonald, S.; Martin, R.; Atkinson, G.J. A Permanent Magnet Assisted Switched Reluctance Machine for More Electric Aircraft. In Proceedings of the International Conference on Electrical Machines (ICEM), Lausanne, Switzerland, 4–7 September 2016.
27. Hwang, H.; Hur, J.; Lee, C. Novel permanent-magnet-assisted switched reluctance motor (I): Concept, design, and analysis. In Proceedings of the International Conference on Electrical Machines and Systems (ICEMS), Busan, Korea, 23–29 October 2013.
28. Lobo, N.S.; Lim, H.S.; Krishnan, R. Comparison of Linear Switched Reluctance Machines for Vertical Propulsion Application: Analysis, Design, and Experimental Correlation. *IEEE Trans. Ind. Appl.* **2008**, *44*, 1134–1142. [CrossRef]
29. Lin, J.; Cheng, K.W.E.; Zhang, Z.; Cheung, N.C.; Xue, X. Adaptive sliding mode technique-based electromagnetic suspension system with linear switched reluctance actuator. *IET Electr. Power Appl.* **2015**, *9*, 50–59. [CrossRef]
30. Chen, Y.; Cao, M.; Ma, C.; Feng, Z. Design and Research of Double-Sided Linear Switched Reluctance Generator for Wave Energy Conversion. *Appl. Sci.* **2018**, *8*, 1700. [CrossRef]
31. Hor, P.J.; Zhu, Z.Q.; Howe, D.; Rees-Jones, J. Minimization of cogging force in a linear permanent magnet motor. *IEEE Trans. Magn.* **1998**, *34*, 3544–3547. [CrossRef]

32. Zhu, Z.Q.; Hor, P.J.; Howe, D.; Rees-Jones, J. Calculation of cogging force in novel slotted linear tubular brushless permanent magnet motor. *IEEE Trans. Magn.* **1997**, *33*, 4098–4100. [CrossRef]
33. Jung, S.Y.; Jung, H.K. Reduction of force ripples in permanent magnet linear synchronous motor. In Proceedings of the International Conference on Electric Machines (ICEM), Bruges, Belgium, 26 August 2002.
34. Kang, G.-H.; Hong, J.-P.; Kim, G.-T. A novel design of an air-core type permanent magnet linear brushless motor by space harmonics field analysis. *IEEE Trans. Magn.* **2001**, *37*, 3732–3736. [CrossRef]
35. Fujii, N.; Okinaga, K. X-Y linear synchronous motors without force ripple and core loss for precision two-dimensional drives. *IEEE Trans. Magn.* **2002**, *38*, 3273–3275. [CrossRef]
36. Liu, X.; Gao, J.; Huang, S.; Lu, K. Magnetic Field and Thrust Analysis of the U-Channel Air-Core Permanent Magnet Linear Synchronous Motor. *IEEE Trans. Magn.* **2017**, *53*, 1–4. [CrossRef]
37. Garcia-Amorós, J. Linear hybrid reluctance motor with high-density force. *Energies* **2018**, *11*, 2805. [CrossRef]
38. Inoue, M.; Sato, K. An approach to a suitable stator length for minimizing the detent force of permanent magnet linear synchronous motors. *IEEE Trans. Magn.* **2000**, *36*, 1890–1893. [CrossRef]
39. Meeker, D.C. Finite Element Method Magnetics, Version 4.2 (28 February 2018 Build). Available online: http://www.femm.info (accessed on 16 October 2018).
40. Altair Flux 3D. Altair 2019. Available online: http://www.altair.com/flux (accessed on 1 October 2020).
41. Zhu, Y.; Lee, S.; Chung, K.; Cho, Y. Investigation of Auxiliary Poles Design Criteria on Reduction of End Effect of Detent Force for PMLSM. *IEEE Trans. Magn.* **2009**, *45*, 2863–2866. [CrossRef]

Article

A Method to Improve Torque Density in a Flux-Switching Permanent Magnet Machine

Junshuai Cao [1], Xinhua Guo [1,*], Weinong Fu [2], Rongkun Wang [1], Yulong Liu [1] and Liaoyuan Lin [1]

[1]   College of Information Science and Engineering, Huaqiao University, Xiamen 361021, China; caojunshuai1995@126.com (J.C.); wangrongkun@hqu.edu.cn (R.W.); yulongliu@hqu.edu.cn (Y.L.); linliaoyuan@hqu.edu.cn (L.L.)

[2]   Department of Electrical Engineering, The Hong Kong Polytechnic University, Hong Kong 999077, China; eewnfu@polyu.edu.hk

*   Correspondence: guoxinhua@hqu.edu.cn

Received: 9 August 2020; Accepted: 10 October 2020; Published: 13 October 2020

Abstract: With the continuous development of machines, various structures emerge endlessly. In this paper, a novel 6-stator-coils/17-rotor-teeth (6/17) E-shaped stator tooth flux switching permanent magnet (FSPM) machine is introduced, which has magnets added in the dummy slots of the stator teeth. This proposed machine is parametrically designed and then compared with the conventional 6/17 E-shaped stator tooth FSPM machine through finite element method (FEM) analysis. Then, combined with the results of FEM, the performance of two machines is evaluated, such as electromagnetic torque, efficiency, back electromotive force (back-EMF). The final results show that this novel 6/17 FSPM machine has greater output torque and smaller torque ripple.

Keywords: flux switching; performance comparison; torque density; permanent magnet (PM)

1. Introduction

Flux switching permanent magnet (FSPM) machines have gained wide application from aerospace to automobile industries since they offer several key advantages, such as a simple and robust rotor, short end winding, high torque density, high efficiency, and excellent flux-weakening capability. The flux switching permanent magnet machine has characteristics of a permanent magnet synchronous machine and a switched reluctance machine, which combines the merits of both. These advantages are particularly important for applications such as electric vehicles, wind power technologies, and flywheel systems. However, compared with the traditional motor, the torque density of a FSPM is low. In this paper, the existing structure is improved to enhance the torque density of a FSPM. After putting forward the operation principle of the FSPM machine, many kinds of its topologies have been studied [1–6]. Among them, the E-shaped stator tooth FSPM machine has a unique structure with two magnets mounted in each stator tooth, which makes it exhibit higher torque/magnetic ratio and larger torque [7].

In this paper, a novel 6/17 E-shaped stator tooth FSPM machine is studied for further improving the torque density. Compared with the conventional 6/17 E-shaped stator tooth FSPM machine, the proposed machine adds permanent magnets in the dummy slot of the stator teeth. In Section 3, the electromagnetic performance of the two machines is analyzed using the finite element method (FEM). Afterwards, the performance of the two machines is compared, according to back-EMF, electromagnetic torque, torque ripple, etc. Finally, the conclusions are drawn in Section 4.

2. Model of Machines

The structure of the conventional 6/17 E-shaped stator tooth FSPM machine can be seen in Figure 1. A dummy slot is inserted between two magnets that are embedded at the upper apex of a stator tooth, which can increase the torque of this machine and reduce torque ripple. However, it is worth noting that for the machine shown in Figure 1, dummy slots are empty. If permanent magnets are added into the dummy slots, the torque will be further improved. In addition, the armature coils of this structure are arranged in the form of A1–B1–C1–A2–B2–C2, and a phase winding is formed by every two coils connected in series, that is, A1 and A2 represent phase A [7]. Moreover, these coils are distributed radially along the space. At the same time, the magnetic polarities formed by magnets near the air gap maintain N–S–S–N.

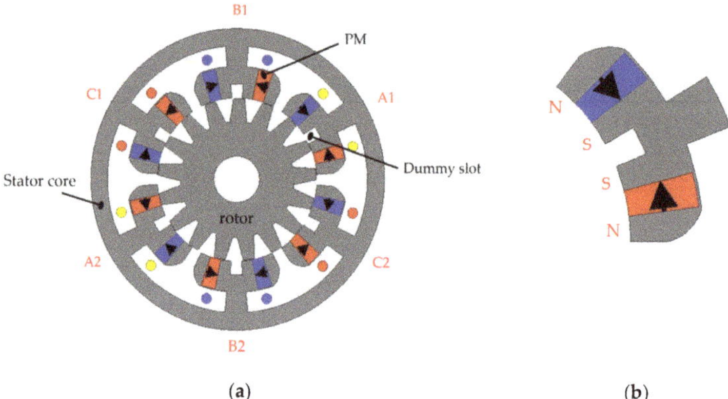

Figure 1. Conventional 6/17 E-shaped stator tooth flux switching permanent magnet (FSPM) machine: (**a**) conventional 6/17 FSPM machine; (**b**) partial enlarged.

As seen in Figure 2, a permanent magnet (PM) is added to each dummy slot of the stator tooth in the proposed structure, so that its effective PM volume increases. Therefore, compared with the initial structure, the output torque of this novel 6/17 E-shaped stator tooth machine is greater.

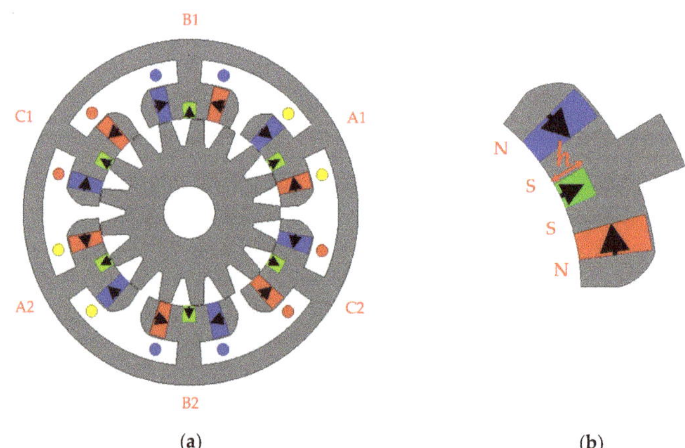

Figure 2. Proposed 6/17 E-shaped stator tooth FSPM machine: (**a**) proposed 6/17 FSPM machine; (**b**) partial enlarged.

For further contrasting the performance of the machine, the motor structure proposed in [7] is reproduced in this paper. Through ANSYS software modeling and parametric analysis, the conventional 6/17 E-shaped stator tooth FSPM machine with the same structure and performance as the machine proposed in [7] is obtained.

This parametric design process mainly focuses on the shape and size of initial permanent magnets, the width and depth of the opening of dummy slots, and the shape of rotor teeth; the optimization results are acquired by using ANSYS software. Then, their maximum average output torque is obtained under the same copper loss (69 w). If the end winding is neglected, copper loss will satisfied the following equation [8]:

$$P_{Cu} = 3I_a^2 R_a = \frac{6I_a^2 N_a^2 \rho_{Cu} L_a}{S_a k_{pf}} \tag{1}$$

where N_a is the number of coil turns per phase, R_a is the phase resistance, L_a is the stack length, S_a is the stator slot area, k_{pf} is the winding packing factor, I_a is the RMS phase current, and ρ_{Cu} is the electrical resistivity of copper at 20 °C.

To analyze the performance of the two 6/17 E-shaped stator tooth FSPM machines, FEM is carried out by using ANSYS software. Based on the initial structure, permanent magnets with radial magnetization are added in the dummy slot. The thickness of the newly added permanent magnet, h in Figure 2, is parametrically designed and simulated.

As shown in Figure 3, when the current density is 5 A/mm^2, average torque and torque ripple of the proposed machine vary with the thickness of the permanent magnet. Figure 3a shows that the torque of the proposed machine is positively proportional to h, and the inflection point is around 3 mm. After 3 mm, the thickness of permanent magnet increases, but the average torque almost does not change. It can be observed from Figure 3a that torque ripple is inversely proportional to h, basically. Meanwhile, the inflection point is at 4 mm, and the torque ripple does not change after 4 mm; thus, the final value of h is 4 mm. In this case, the average torque is the largest and torque ripple is the smallest, while the amount of permanent magnets is the least. Excessive use of permanent magnets will increase the cost of manufacture and the iron loss. Table 1 shows their main specifications and parameters.

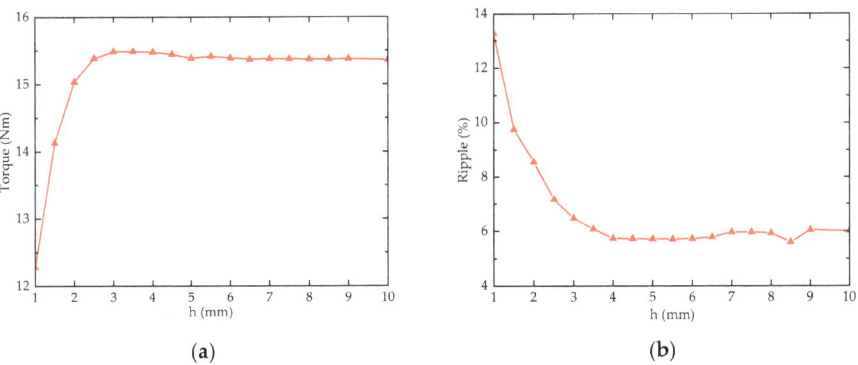

(a) (b)

Figure 3. The results of parameterization: (a) relationship between average torque and h; (b) relationship between torque ripple and h.

Table 1. Design parameters of two FSPM machines.

Items	Initial	Novel
Stator coils number	6	6
Rotor teeth number	17	17
Stack length (mm)	75	75
Stator outer diameter (mm)	130	130
Stator inner diameter (mm)	70	70
Air-gap length (mm)	0.35	0.35
Slot package factor (k_{pf})	0.45	0.45
Machine volume (m^3)	9.95×10^{-4}	9.95×10^{-4}
Rated speed (r/min)	1500	1500
Rated current density (A/mm^2)	5	5
PM type	N35SH	N35SH
PM volume (mm^3)	63,900	77,400
Remnant B_r (T)	1.2	1.2
Coercivity H_c (kA/m)	909	909
Stator slot area (mm^2)	408	408
Coil turns	73	73
Coil number per phase	2	2

3. Performance Comparisons

3.1. No-Load Performance

Meanwhile, it can be distinguished from Figure 4 that the no-load magnetic field distributions of the novel structure are different from the initial structure. Additionally, in the structure of the proposed machine, the magnetic flux density of its stator is somewhat greater than that of the conventional structure. In Figure 4, from the circled part, nine magnetic lines on Figure 4a and 10 lines on Figure 4b can be seen. It can be observed that the flux density in Figure 4b is higher. As shown in Figure 4a, there is no flux distribution in the dummy slot, but, magnetic lines of magnetization direction are added in the dummy slot for Figure 4b. This is because permanent magnets are increased in the dummy slots of the proposed structure. It can be observed from Figure 4 that the distortion of the magnetic lines of force is also reduced with the increased permanent magnet. The higher the magnetic density at the stator, the greater the torque that can be produced under the same current.

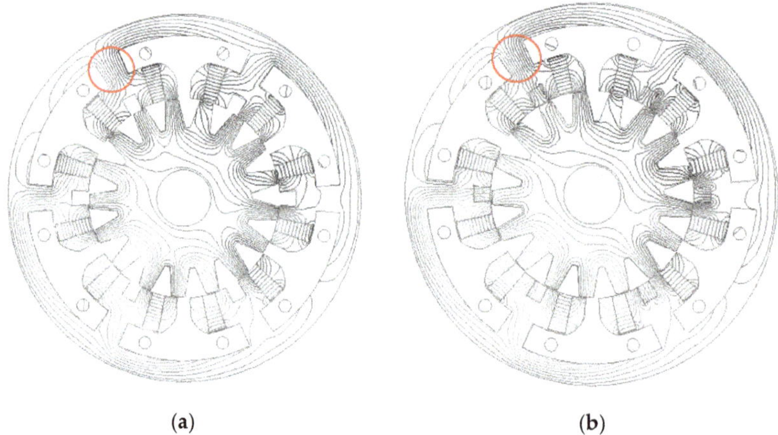

(a)   (b)

Figure 4. No-load magnetic flux distributions of two FSPM machines: (**a**) conventional 6/17 FSPM machine; (**b**) proposed 6/17 FSPM machine.

The distribution of no-load air-gap flux density in two E-shaped stator tooth FSPM machines is implied by Figure 5. It can be observed that the proposed machine has a larger air-gap flux density, which indicates the proposed structure can produce a much larger output torque than the initial one.

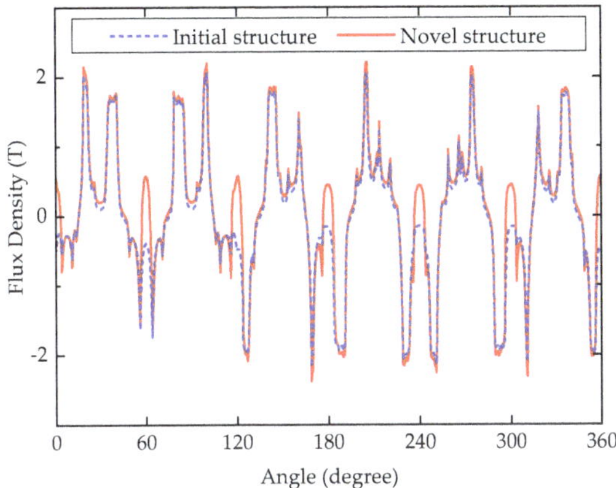

Figure 5. No-load air-gap flux density distributions for the two FSPM machines.

At the same time, peak-to-peak value of cogging torque of the proposed machine (0.54 Nm) is higher than that of the initial machine (0.33 Nm), as shown in Figure 6. Additionally, the energy in the air gap is related to cogging torque, so the formula for calculating the cogging torque [9] can be written as:

$$T_{cog} = -\frac{\partial W}{\partial \theta_r} = -\frac{\partial}{\partial \theta_r}\left(\frac{1}{2\mu_0}\int B^2 dV\right) \tag{2}$$

where W is the energy in the air gap, B is the air-gap flux density, μ_0 is the magnetic permeability of the free space, and V is the volume of air gap. Meanwhile, this novel 6/17 E-shaped stator tooth FSPM has both a higher air-gap flux density and a larger cogging torque due to the magnets in the dummy slots of the stator tooth.

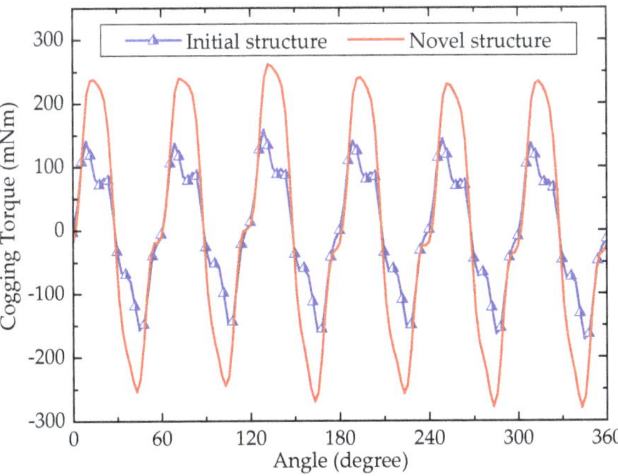

Figure 6. Cogging torque waveforms of the two FSPM machines.

Next, Figure 7 shows the no-load back-EMF waveforms of the two machines at 1500 r/min that were computed by finite element method (FEM). Results indicate that the two waveforms are symmetrical; in addition, with respect to the amplitude back-EMF, the proposed E-shaped stator tooth machine shows a larger amplitude back-EMF because of larger fluxes flowing through the coil. Meanwhile, Figure 8 indicates the fast Fourier transform (FFT) results of back-EMF waveforms.

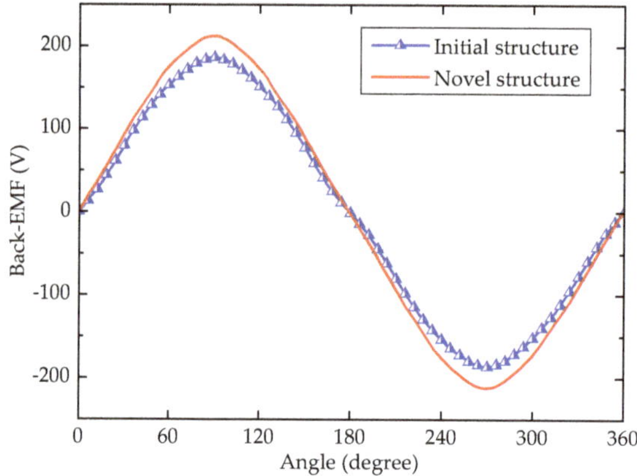

Figure 7. No-load back-EMFs of the two FSPM machines.

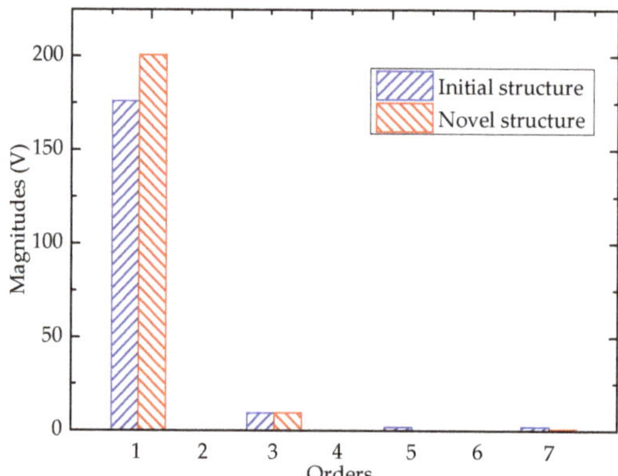

Figure 8. Fast Fourier transform (FFT) results of back-EMFs of the two FSPM machines.

After adding magnets in the dummy slots of the stator teeth, the amplitude of fundamental back-EMFs is enhanced. It can be observed from Figure 8 that the third, fifth, and seventh components harmonics of on-load back-EMF are reduced greatly after adding magnets. Then, with magnets located in the dummy slots, the fundamental harmonic of on-load back-EMF is enhanced from 176.1 to 200.6 V. The total harmonic distribution (THD) of initial and proposed E-shaped stator tooth machines is 5.7% and 4.7%, respectively, and there is no additional extra harmonic introduced into the targeted on-load back-EMF.

3.2. Torque Performance

Figures 9 and 10 indicate that the two machines are controlled by id = 0 mode, where the current density ranges from 1 to 10 A/mm^2, while the stator coil turns and the slot packing factor is 73 and 0.45, respectively [8]. Results imply that under 10 A/mm^2, the average torque of the proposed structure is significantly greater than that of the initial structure. As far as torque ripple is concerned, the proposed machine is also somewhat smaller than the E-shaped stator tooth structure above 2 A/mm^2, after adding the magnets in the dummy slots of stator teeth. The average torque of the proposed machine (15.38 Nm) is greater than that of the conventional one (13.04 Nm) at the rated current density (5 A/mm^2).

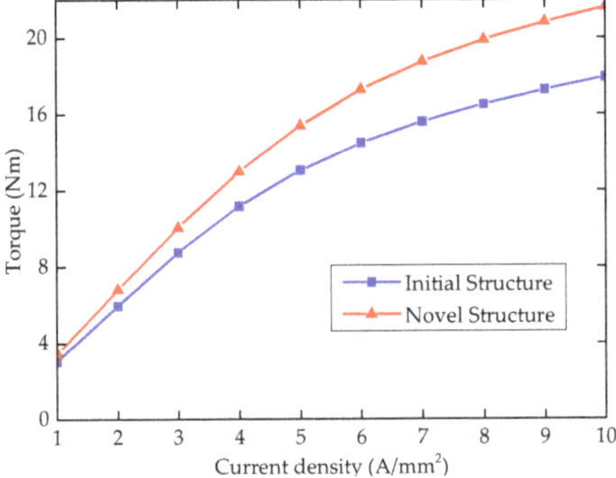

Figure 9. Average torque waveforms versus the current density.

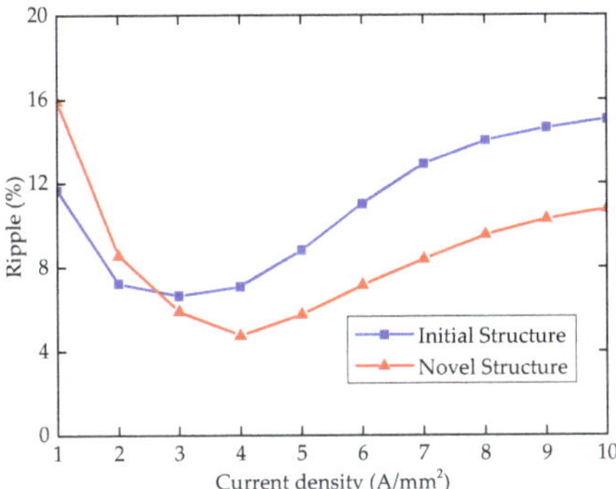

Figure 10. Torque ripple waveforms versus the current density.

The torque ripple of the novel machine (5.7%), however, is a little lower. Under the condition of low torque ripple, the proposed machine needs to be further optimized through adopting some strategies to make the torque ripple smaller [5,10,11]; at the same time, the average torque will also be reduced. Figure 11 indicates the electromagnetic torque waveforms of these two machines.

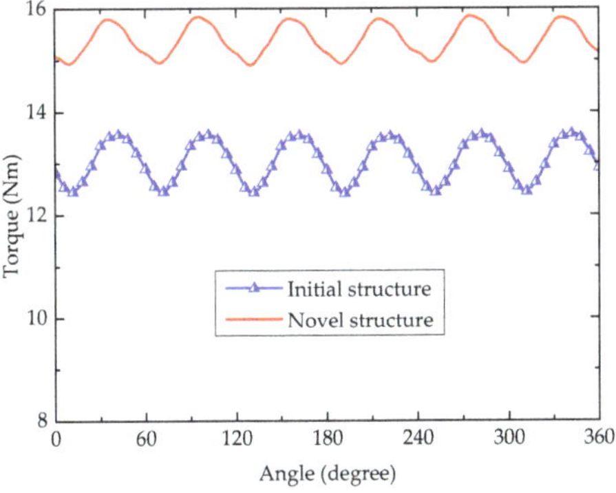

Figure 11. Electromagnetic torque waveforms under J = 5 A/mm^2.

Table 2 shows comparative results of the two E-shaped stator tooth FSPM machines. It can be found that they both have the same machine volume, but the ratio of torque to the volume of the proposed machine is larger. More importantly, the average torque of the machines is 13.04 Nm and 15.38 Nm, respectively; thus, the average torque of the proposed machine is 20% higher. In other words, torque density is hugely enhanced for the proposed machine. Meanwhile, the torque ripple of the two machines is 5.7% and 8.8%, respectively, and that of the proposed machine is reduced by 3.1%. This is because the new structure increases the radial permanent magnet and enhances flux density distribution in the air gap, so the cogging torque and torque ripple are reduced. Besides, when the two machines both work under 1500 r/min and 5 A/mm^2, their efficiency is approximately equal, 0.877 and 0.897. The amount of permanent magnets increases by 21%, which improves the performance of the motor, and the cost of permanent magnet is only a small part of the manufacturing cost of the motor. Therefore, it is worthwhile to increase the permanent magnets.

Table 2. Comparative results of the two FSPM machines.

Items	Initial	Novel
Amplitude of fundamental phase back-EMF (V)	176.1	200.6
Total harmonic distribution (THD) of phase back-EMF	5.7%	4.7%
Average torque (Nm)	13.04	15.38
Torque ripple	8.8%	5.7%
Ratio of torque to machine volume (Nm/m^3)	13,169	15,530
Efficiency at 5 A/mm^2 and 1500 r/min	0.877	0.897

4. Conclusions

In summary, this paper introduces a novel 6/17 E-shaped stator tooth FSPM machine with an added magnet in the upper apex of each dummy slot, which is derived from the conventional 6/17 E-shaped stator tooth FSPM machine. Then, two kinds of 6/17 E-shaped stator tooth FSPM machine topologies with the same rotor tooth are compared by FEM. Finally, the electromagnetic performance of the two kinds of E-shaped stator tooth machine is compared, while final results indicate that the proposed E-shaped stator tooth machine has higher torque, back-EMF, and efficiency than the initial machine. In addition, the proposed structure has a lower torque ripple and lower THD of phase back-EMF. Therefore, this proposed E-shaped stator tooth machine has a better performance.

Author Contributions: Conceptualization, J.C. and X.G.; methodology, W.F.; software, J.C.; validation, J.C., Y.L. and X.G.; formal analysis, J.C.; investigation, J.C.; resources, X.G.; data curation, J.C.; writing—original draft preparation, R.W.; writing—review and editing, L.L.; visualization, J.C.; supervision, W.F.; project administration, X.G.; funding acquisition, X.G. All authors have read and agreed to the published version of the manuscript.

Funding: This work was supported by the Xiamen Science and Technology Project—University Research Institute Industry–University–Research Project—under Grant 3502Z202003037, and in part by the Subsidized Project for Postgraduates' Innovative Fund in Scientific Research of Huaqiao University under Grant 18014082018.

Conflicts of Interest: The authors declare no conflict of interest.

References

1. Hoang, E.; Ahmed, B.; Lucidarme, J. Switching flux permanent magnet polyphased synchronous machines. In Proceedings of the European Conference Power Electrinics and Applications, Trondheim, Norway, 8–10 September 1997; pp. 903–908.
2. Wu, Z.Z.; Zhu, Z.Q. Analysis of air-gap field modulation and magnetic gearing effects in switched flux permanent magnet machines. *IEEE Trans. Magn.* **2015**, *51*, 1–12. [CrossRef]
3. Hua, W.; Cheng, M.; Zhu, Z.Q.; Howe, D. Analysis and optimization of back EMF waveform of a flux-switching permanent magnet motor. *IEEE Trans. Energy Convers.* **2008**, *23*, 727–733. [CrossRef]
4. Chen, J.T.; Zhu, Z.Q. Winding configurations and optimal stator and rotor pole combination of flux-switching PM brushless AC machines. *IEEE Trans. Energy Convers.* **2010**, *25*, 293–302. [CrossRef]

5.   Xiang, Z.; Zhu, X.; Quan, L.; Du, Y.; Zhang, C.; Fan, D. Multilevel design optimization and operation of a brushless double mechanical port flux-switching permanent-magnet motor. *IEEE Trans. Ind. Electron.* **2016**, *63*, 6042–6054. [CrossRef]

6.   Zhou, Y.J.; Zhu, Z.Q. Torque density and magnet usage efficiency enhancement of sandwiched switched flux permanent magnet machines using V-shaped magnets. *IEEE Trans. Magn.* **2013**, *49*, 3834–3837. [CrossRef]

7.   Zhao, G.; Hua, W. Comparative study between a novel multi-tooth and a V-shaped flux-switching permanent magnet machines. *IEEE Trans. Magn.* **2019**, *55*, 1–8. [CrossRef]

8.   Guo, X.; Wu, S.; Fu, W.N.; Liu, Y.; Wang, Y.; Zeng, P. Control of a dual-stator flux-modulated motor for electric vehicles. *Energies* **2016**, *9*, 517. [CrossRef]

9.   Guo, X.; Wang, Q.; Shang, R.; Chen, F.; Fu, W.; Hua, W. Design and analysis of a novel synthetic slot dual-PM machine. *IEEE Access* **2019**, *7*, 29916–29923. [CrossRef]

10. Zhu, X.; Hua, W.; Cheng, M. Cogging torque minimization in flux-switching permanent magnet machines by tooth chamfering. In Proceedings of the IEEE Energy Conversion Congress and Exposition, Milwaukee, WI, USA, 18–22 September 2016; pp. 1–7.

11. Zhu, X.; Hua, W.; Wu, Z.; Huang, W.; Zhang, H.; Cheng, M. Analytical approach for cogging torque reduction in flux-switching permanent magnet machines based on magnetomotive force-permeance model. *IEEE Trans. Ind. Electron.* **2018**, *65*, 1965–1979. [CrossRef]

Article

PV Generator-Fed Water Pumping System Based on a SRM with a Multilevel Fault-Tolerant Converter

Vitor Fernão Pires [1,2], Daniel Foito [1,3], Armando Cordeiro [1,2,4], Miguel Chaves [2,4] and Armando J. Pires [1,3,*]

1 SustainRD, EST Setubal, Polytechnic Institute of Setúbal, 2914-508 Setúbal, Portugal;
   vitor.pires@estsetubal.ips.pt (V.F.P.); daniel.foito@estsetubal.ips.pt (D.F.); armando.cordeiro@isel.pt (A.C.)
2 Instituto de Engenharia de Sistemas e Computadores—Investigação e Desenvolvimento em
   Lisboa (INESC-ID), 1000-029 Lisboa, Portugal; miguel.chaves@isel.pt
3 Centre of Technology and Systems (CTS-UNINOVA), 2829-516 Caparica, Portugal
4 ISEL—Instituto Politécnico de Lisboa, 1959-007 Lisboa, Portugal
* Correspondence: armando.pires@estsetubal.ips.pt

Abstract: This paper presents a pumping system supplied by a PV generator that is based on a switched reluctance machine (SRM). Water pumping systems are fundamental in many applications. Most of them can be used only during the day; therefore, they are highly recommended for use with PV generators. For the interface between the PV panels and the motor, a new multilevel converter is proposed. This converter is designed in order to ensure fault-tolerant capability for open switch faults. The converter is based on two three-level inverters, with some extra switches. Moreover, to reduce the number of switches, the converter is designed to provide inverse currents in the motor windings. Due to the characteristics of this motor, the inverse currents do not change the torque direction. In this way, it was possible to obtain an SRM drive with fault-tolerant capability for transistor faults; it is also a low-cost solution, due to the reduced number of switches and drives. These characteristics of fault-tolerant capability and low cost are important in applications such as water pumping systems supplied by PV generators. The proposed system was verified by several tests that were carried out by a simulation program. The experimental results, obtained from a laboratory prototype, are also presented, with the purpose of validating the simulation tests.

Keywords: SRM; PV panels; water pumping system; multilevel converter; fault-tolerant

Citation: Pires, V.F.; Foito, D.; Cordeiro, A.; Chaves, M.; Pires, A.J. PV Generator-Fed Water Pumping System Based on a SRM with a Multilevel Fault-Tolerant Converter. *Energies* **2022**, *15*, 720. https://doi.org/10.3390/en15030720

Academic Editor: Thanikanti Sudhakar Babu

Received: 23 December 2021
Accepted: 13 January 2022
Published: 19 January 2022

Publisher's Note: MDPI stays neutral with regard to jurisdictional claims in published maps and institutional affiliations.

1. Introduction

One of the applications for which PV generators are considered highly adequate is water pumping systems [1–4]. In many of these systems the operation is compatible with the intermittency of the electrical energy supplied by the generators. Therefore, it is possible to implement a completely renewable and clean solution that does not produce greenhouse gases. On the other hand, it is extremely important and appropriate to implement it in rural and remote areas [4–6]. Taking these factors into consideration, the development of a reliable and economical pumping system is fundamental.

One of the aspects that must be considered in a pumping system is the choice of the electric motor. Several types of motor can be used in this type of application [6]. However, one of the motors that is considered the most interesting is the switching reluctance machine (SRM). When compared to other industrial motors, this motor is considered simple in design [7]. As a consequence, several applications and research studies have been performed [8–12]. From these applications and studies, it was possible to verify that in fact, the SRM showed excellent characteristics for this kind of application.

One fundamental aspect of the application of an SRM is the electronic power converter that must be used to operate the motor; several topologies can be adopted. The classic solution is the asymmetrical half-bridge topology [13–15]. However, many other two-level topologies have been presented and proposed. Most of those topologies are characterized to reduce the number of electronic power components [16–20]. However, electronic power converters are prone to faults in their switches [21]. Thus, in order to consider this aspect, several topologies have been developed with the purpose of providing fault-tolerant capability for this type of faults. The approach to developing these converters is based on the introduction of extra switches and/or relays. An example in which an extra leg and relays to connect it to the faulty phase are introduced can be seen in [22]. Another topology that introduces new switches and relays was also proposed in [23]. In this topology, instead of using a leg with a switch and a diode, a classical leg with two switches and two diodes is used. One interesting concept that is used in the context of this topology is that in fault-tolerant mode, changes in the current excitation are also considered. Another topology in which extra switches were introduced, but with the purpose of minimizing them, is presented in [24]. A fault-tolerant system with extra switches and windings (dual-channel switched reluctance motor) was also proposed by [25]. It is characterized by two operational models. The first operational model works like a three-phase conventional SRM delivered by an H-bridge inverter, driven by square-wave currents. The second operational model involves two SRMs, mutually coupled, being supplied by sine-wave currents. These operational models are used to solve various faults, and the corresponding remedial current strategies are proposed to perform fault-tolerant operation. In [26], another topology was proposed with extra switches that can be arranged to work in the usual and modular driving modes so that the advantages of both driving methods can be applied. A topology with two additional switches and six thyristors, complementing the classical AHB, was also proposed [27]. In [28], a topology that uses a standby single-phase full bridge in combination with an extra group of mechanical switches replacing up to four faulty active switches was presented. However, this solution is not capable of isolating short-circuit failures in controlled-power semiconductors. The fault-tolerant solution presented in [29] combines a classical AHB topology with a classical redundant VSI topology applied to a three-phase SRM. The redundant three-phase VSI is linked to every central-tapped winding phase. In [30], a fault-tolerant topology with an additional classical single-phase full-bridge joined with six mechanical relays applied to a three-phase SRM was presented. Each phase of the converter is divided into three windings to connect additional controlled power semiconductors in fault-tolerant operation. In [31], the use of two inverter legs for each phase for a special six-phase 12/8 SRM was proposed to increase drive system reliability, based on the concept of increasing the number of legs per phase of the inverter. This proposal uses a mutually coupled dual three-phase SRM.

Solutions in which only relays are used were also proposed [32,33]. The first solution benefits from the independence between SRM phases and demands static switches linked between phases that are not in sequence. The second solution uses the same principle presented in [32] with some enhancements regarding fault-tolerant operation. This solution uses healthy legs to control two phases that are not in a sequence; for example, in a four-phase SRM, phases 'A' and 'C' or phases 'B' and 'D'. It uses various changeover solid-state relays (SSRs) to link the faulty legs to other healthy phases, so as to replace and isolate the faulty devices. This solution offers the advantage of isolating faulty controlled power semiconductors regardless of the failure mode. Another category of power converter topology that can be used with the SRM features the ability to generate multilevel voltages. Consequently, it is possible to improve the performance of the machine and reduce the power semiconductors' switching frequency [34]. Another important aspect of multilevel topologies is that they can provide fault-tolerant capability. In this way, T-type and neutral point clamped asymmetric half-bridge (NPC) topologies have been studied [34–38]. However, these topologies present some limitations regarding their fault-tolerant capability for all types of faults. Topologies based on the modular multilevel converter (MMC) were also proposed. One of the solutions is the classical asymmetric MMC [34]. In this solution, the MMC's structure is linked to the asymmetric bridge converter [39]. In [40], an NPC-AHB converter with intrinsic dc-link voltage boosting capacitors was presented. This topology was developed for an 8/6 SRM and characterized by two NPC-AHB converters, one of which was connected to phases A and C and the other to phases B and D. This converter was proposed for applications involving high-speed electric vehicles. In [41], a modular structure based on an asymmetric half-bridge converter topology, with a central-tapped winding node, for an 8/6 SRM, was proposed. This solution was specifically developed for electric and hybrid electric vehicles. Another proposed topology [42] also offers boost capability, but with symmetrical legs and a common point for the machine windings. This topology features a high number of power switches and requires complex control to balance the floating capacitor.

Topologies based on these classical forms, but with reduced numbers of switches, were also proposed [43–46]. However, they become severely affected regarding their fault tolerance capability. In this way, an NPC topology, modified by adding active switches to the inverter's clamping diodes, combined with another group of power switches and diodes linked to each branch of the converter, was also proposed [47]. However, the number of switches of the converter is very high. Thus, to provide a complete fault tolerance capability regarding all types of power semiconductors faults, and even multiple faults with a reduced number of switches, a four-quadrant NPC topology for an SRM was proposed [48]. This topology also uses the concept that in fault-tolerant mode, the change in the current excitation is also considered. However, this solution is still characterized by a high number of controlled-power semiconductors.

With the objective of providing multilevel operation, as well as fault-tolerant capability for open-switch faults, a new topology is proposed in this work. Few studies have addressed multilevel converters to SRMs with fault-tolerant capability. Thus, the proposed solution provides multilevel characteristics, but contrary to the solutions described previously, this topology is characterized by a reduced number of power semiconductors. This results in an SRM with fault-tolerant capability for open-switch faults. Thus, with the reduction in the number of switches and transistor drives, it is possible to reduce the cost of the solution. These aspects of fault-tolerant capability and reduced costs are extremely important in several applications in water pumping systems supplied by PV generators. In reality, these applications are often used in poor countries and in remote regions. Furthermore, in fault-tolerant mode, the change in the current excitation is also considered. The characteristics and fault-tolerant capability of the proposed topology are verified through several tests using a simulation platform and a laboratory prototype.

2. Torque Characteristic of the SRM

In an SRM, the reluctance of the magnetic circuit depends on the position of the rotor. The reluctance circuit is influenced by the geometry of the machine and some of its constructive parameters, such as the type, thickness, and lamination factor of the used ferromagnetic material. To apply a (reluctance) torque in this machine, a variation of the reluctance is fundamental, and controlling the time of energizing and deenergizing of the stator phases is necessary to adequately control it. The mathematical model of an SRM is considered complex, with nonlinearities in the magnetic circuit. The SRM electrical equation can be expressed as Equation (1), for each phase. Usually, the mutual inductance is neglected, which means that the magnetic influence among phases is not considered:

$$u_j = R_j i_j + \frac{d\psi_j(\theta_r, i_j)}{dt} \tag{1}$$

where u_j is the phase voltage, i_j the phase current, R_j the phase resistance, θ_r the rotor position, j is the considered phase, and $\psi_j(\theta_r, i_j)$ is the phase linkage flux. Equation (1) can be rewritten as:

$$u_j = R_j i_j + \frac{\partial \psi_j(\theta_r, i_j)}{\partial i_j}\frac{di_j}{dt} + \frac{\partial \psi_j(\theta_r, i_j)}{\partial \theta_r}\omega_r \tag{2}$$

In Equation (2), the last term corresponds to the back EMF, which depends on the speed of the rotor, ω_r, meaning that the back EMF of the motor will present high values during high-speed operation. Consequently, the input voltage should be even higher to guarantee the phase current. For this situation, a multilevel converter could be the adequate solution, imposing a high value of DC voltage. Particularly during periods of phase energizing and deenergizing, high voltage values are required to minimize the commutation time between phases and the possibility of negative values for phase torque. The multilevel converter could also be advantageous at lower speeds, using lower voltage values, improving the efficiency level by diminishing the switching frequency.

To complete the mathematical model of the SRM, an equation for the torque developed by each phase (T_j) should be written. To obtain the total torque, the torque developed by each phase should be added for each time instant. Each phase torque is obtained by calculating the variation of the magnetic co-energy (W_C) related to the variation of the rotor position, as in Equation (3):

$$T_j(\theta_r, i_j) = \left. \frac{\partial W_C(\theta_r, i_j)}{\partial \theta_r} \right|_{i_j=const} \tag{3}$$

where the magnetic co-energy is defined in Equation (4):

$$W_C(\theta_r, i_j) = \int_0^{i_j} \psi_j(\theta_r, i)\, di \tag{4}$$

The linkage flux of each phase can be expressed as in Equation (5):

$$\psi_j(\theta_r, i_j) = L_j(\theta_r) i_j \tag{5}$$

where the magnetic self-inductance coefficient (L_j) is introduced, not depending on the phase current, i_j. However, it could be considered non-linear, depending on the position of the rotor [48].

Using Equation (5), the torque developed by each phase and represented in Equation (3) can be rewritten as:

$$T_j(\theta_r, i_j) = \frac{1}{2} \frac{dL_j(\theta_r)}{d\theta_r} i_j^2 \tag{6}$$

Which leads to the main conclusion, that the torque value is independent of the phase current signal. This conclusion is relevant to sustain the methodology proposed here.

3. Proposed Fault Tolerant Multilevel Converter for a SRM

As mentioned in the first section, SRMs are highly recommended for application to pumping systems. However, these motors require a power converter. On the other hand, if the system is supplied by *PV* panels, a DC–DC converter between those panels and the SRM power converter is usually required. This typical structure is shown in Figure 1, where it is possible to see the three main parts, the PV generators, the power converter, and the SRM.

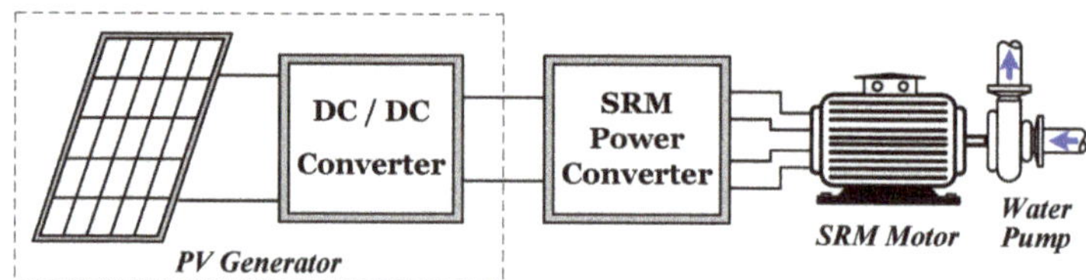

Figure 1. Typical structure of a pumping system with a PV generator and a SRM.

SRMs are typically supplied by two-level power converters. However, as mentioned previously, multilevel converters present several advantages, due to which they are an increasingly interesting alternative. However, these converters do not provide fault-tolerant operation or require a high number of switches. Therefore, a multilevel converter with fault-tolerant operation and a reduced number of switches is proposed in this work. The proposed solution, which is adapted for a 8/6 SRM, is presented in Figure 2. As can be seen, this topology consists of two three-phase two-level voltage source inverters and extra switches. These extra switches are used to connect both positive buses of the three-phase voltage source inverters (bidirectional switch) and, in turn, to connect these to the upper capacitor. The PV generator is connected to the lower capacitor; accordingly, the voltage of the upper capacitor should be controlled.

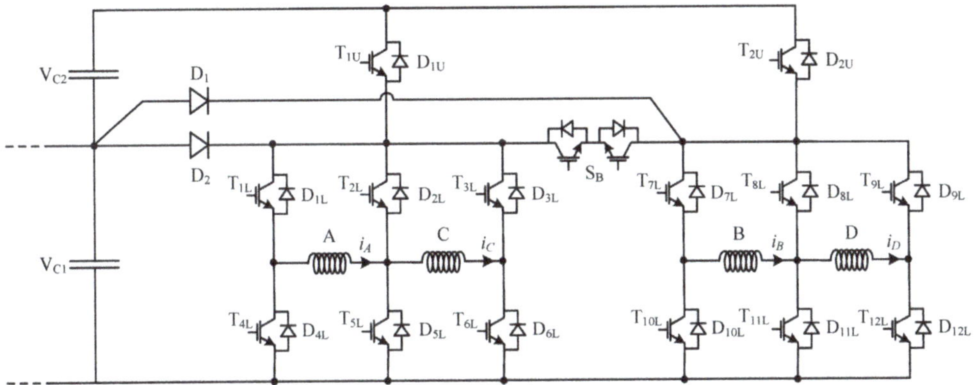

Figure 2. Proposed multilevel fault tolerant converter for the 8/6 SRM.

When the converter operates in healthy mode, the bidirectional switch S_B is always off. This converter allows multilevel operation, since it allows the application of four voltage levels to each motor winding four voltage level, namely, $+V_{DC1} + V_{DC2}$, $+V_{DC1}$, 0 and $-V_{DC1}-V_{DC2}$. Therefore, there are four possible topological arrangements, as shown in Figure 3 for the case of phase A. The application of the maximum positive voltage ($+V_{DC1} + V_{DC2}$) is obtained when the power semiconductors T_{1U}, T_{1L}, and T_{5L} are ON (Figure 3a). The intermediate positive voltage ($+V_{DC1}$) is obtained with the two power semiconductors T_{1L} and T_{5L} in ON state (Figure 3b). Regarding the state associated with the 0 voltage, it is obtained when diode D_{4L} and transistor T_{5L} are ON (Figure 3c). This last state can also be achieved with the diode D_{2L} and transistor T_{1L} in the ON condition. Finally, for the last state (the application of the negative voltage $-V_{DC1}-V_{DC2}$), the diodes D_{2L} and D_{4L} are ON (Figure 3d).

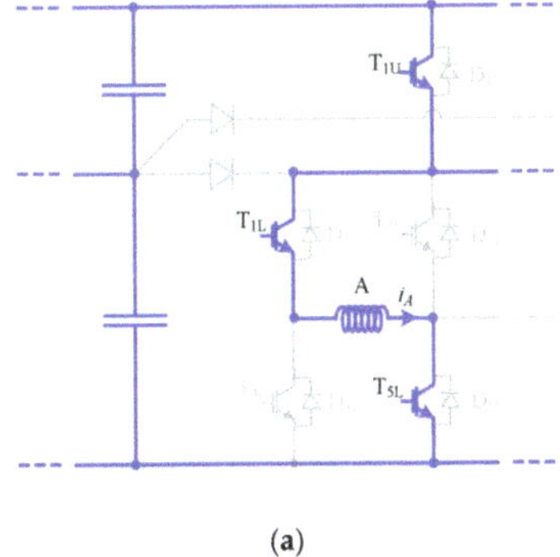

(a)

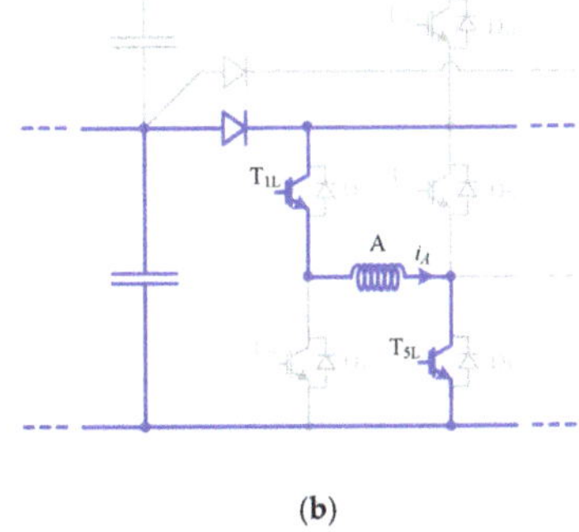

(b)

Figure 3. *Cont.*

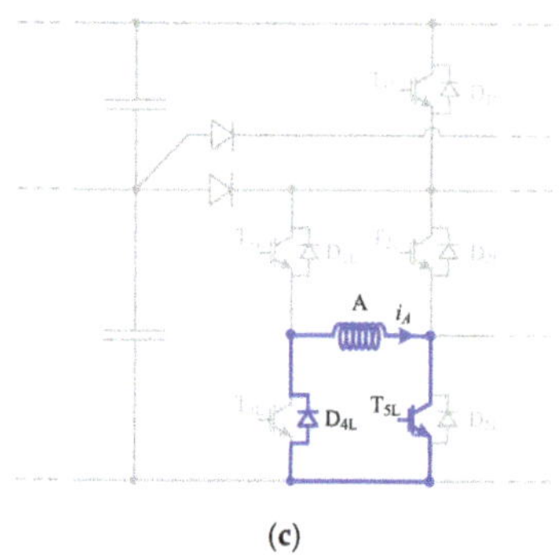

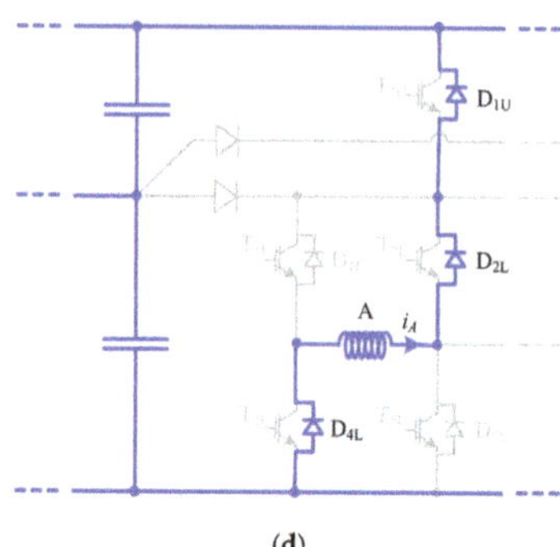

(c)   (d)

Figure 3. Topological arrangements of the multilevel converter associated to phase A in healthy condition: (**a**) voltage level $+V_{C1} + V_{C2}$; (**b**) voltage level $+V_{C1}$; (**c**) voltage level 0; (**d**) voltage level $-V_{C1}-V_{C2}$.

One important characteristic of the proposed topology is that during an open switch fault, it can provide fault-tolerant capability. This fault-tolerant capability is achieved by changing the switches that are controlled under the concept of the bidirectional current excitation. Let us consider the example of a fault in the transistor T_{1L}. During this kind of fault, it is not possible to apply a positive voltage to the motor winding phase A. Thus, to settle this problem, instead of using transistors T_{1L} and T_{5L}, transistors S_{2L} and T_{4L} are used. Consequently, the current in the motor winding phase A flows in the opposite direction when compared with the healthy condition. In Figure 4, it is possible to analyze the new circuits for this fault-tolerant mode. If the fault is in the upper switch, such as T_{1U}, then when the maximum voltage applied to the motor windings A or C is needed, instead of this transistor, transistor T_{2U} and the bidirectional switch S_B should be used. The new circuits for this fault condition can be observed in Figure 5. It should be noted that for this kind of fault, the change in direction of the current excitation is not needed.

Through the analysis that was performed for the proposed topology, it was possible to confirm the system's fault-tolerant capability. As verified, this capability was achieved by changing the transistors that were being used by other ones and in some conditions by the inversion of the current excitation in the winding affected by the fault.

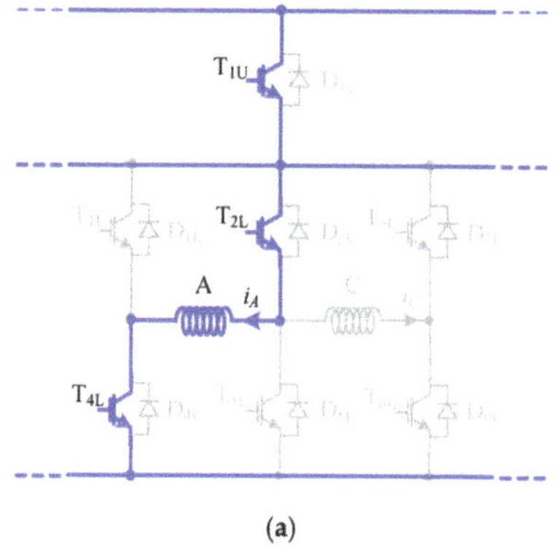

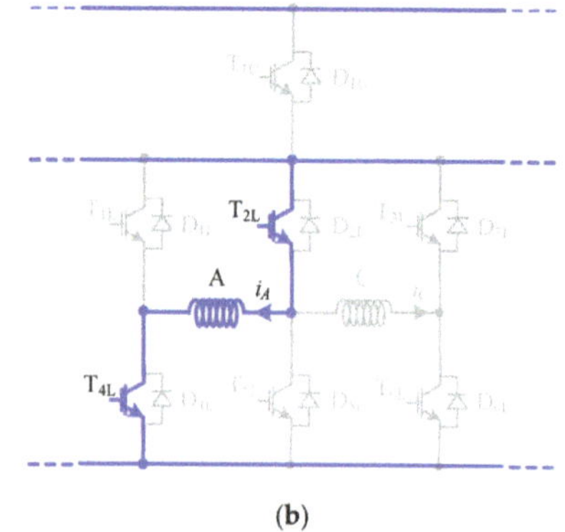

(a)   (b)

Figure 4. *Cont.*

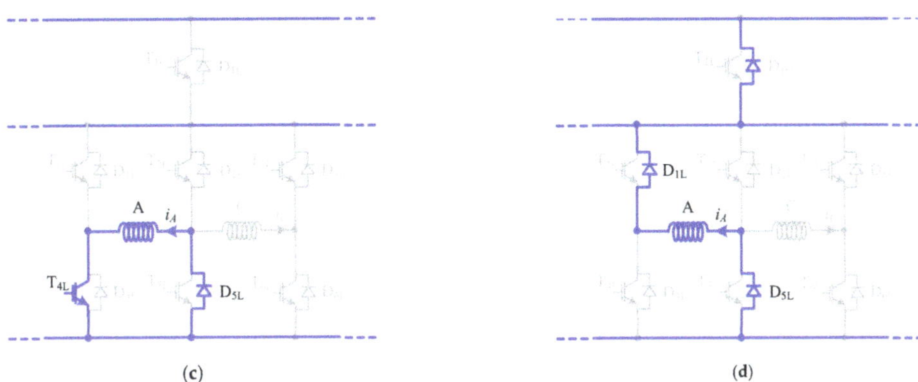

Figure 4. Topological arrangements of the multilevel converter associated to phase A in S_{1L} open transistor fault and fault-tolerant mode: (**a**) voltage level $+V_{C1} + V_{C2}$; (**b**) voltage level $+V_{C1}$; (**c**) voltage level 0; (**d**) voltage level $-V_{C1}-V_{C2}$.

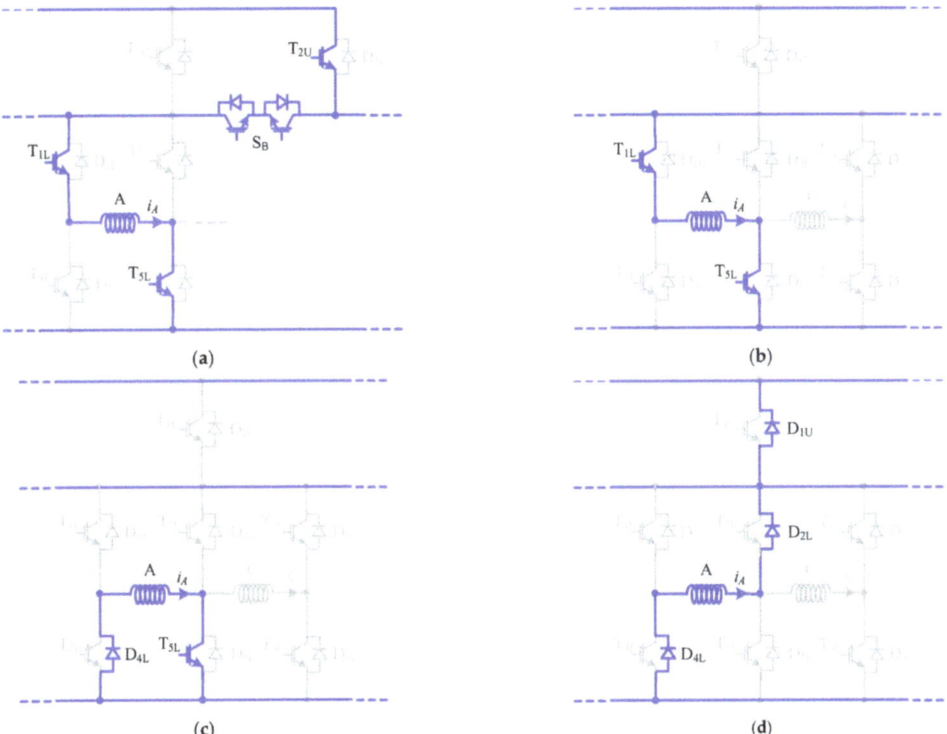

Figure 5. Topological arrangements of the multilevel converter associated to phase A in S_{1U} open transistor fault and fault-tolerant mode: (**a**) voltage level $+V_{C1} + V_{C2}$; (**b**) voltage level $+V_{C1}$; (**c**) voltage level 0; (**d**) voltage level $-V_{C1}-V_{C2}$.

Since the developed topology was designed with the purpose of reducing the number of switches and drives for the transistors, there is a consequence regarding the power switches' balanced losses. In fact, this is the disadvantage of this topology, since the two upper switches, T_{1U} and T_{2U}, will experience greater power losses, since each of them are common to two phases. However, this fact is true for higher motor speeds, since it is in these operational modes that the maximum positive voltage is required. For lower speeds, it is the contrary. In this last case, the maximum voltages are practically not required; accordingly, the upper switches are not operated.

4. Control of the Proposed Converter

To control the proposed multilevel fault-tolerant converter, from the point of view of the motor, several approaches can be used. For this work, a current controller for the motor windings was used. The adopted approach can be seen in Figure 6, where the full scheme is presented. As shown in this figure, the current controller is associated with a modulator that is specially developed for this fault-tolerant converter.

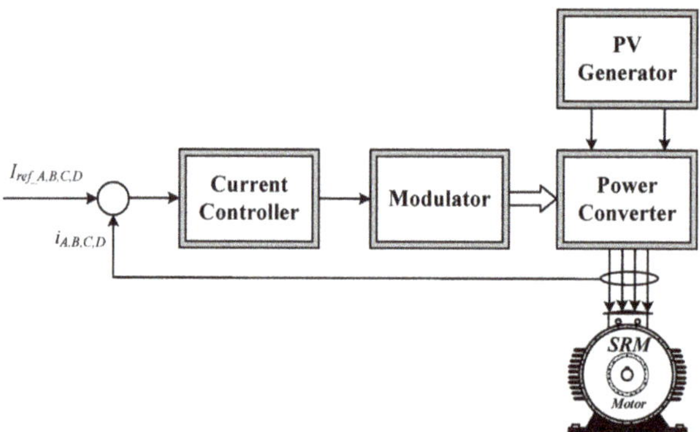

Figure 6. Full scheme of the adopted control approach for the proposed multilevel fault-tolerant converter.

The current controller is responsible for the regulation of the motor winding currents. For this purpose, a current hysteresis controller was used, but designed to allow the generation of the four voltage levels (generating at their output four different values that are associated with the switch combination to generate the required voltage). In this way, the controller followed the conditions presented in Equation (7). The generation of the four different values can also be seen in Figure 7.

$$\begin{cases} if\ i_{ref} - i \geq +2\Delta i & \Rightarrow \lambda = 4 & \Rightarrow V_{phase} = +V_{C1} + V_{C2} \\ if\ +\Delta i \leq i_{ref} - i < +2\Delta i & \Rightarrow \lambda = 3\ or\ 4 & \Rightarrow V_{phase} = +V_{C1}\ or\ +V_{C1} + V_{C2} \\ if\ 0 \leq i_{ref} - i < +\Delta i & \Rightarrow \lambda = 2\ or\ 3 & \Rightarrow V_{phase} = 0\ or\ +V_{C1} \\ if\ -\Delta i \leq i_{ref} - i < 0 & \Rightarrow \lambda = 1\ or\ 2 & \Rightarrow V_{phase} = 0\ or\ -V_{C1} - V_{C2} \\ if\ i_{ref} - i < -\Delta i & \Rightarrow \lambda = 1 & \Rightarrow V_{phase} = -V_{C1} - V_{C2} \end{cases} \tag{7}$$

A modulator is developed to generate the switch combination as a result of the current controller. Therefore, considering the four values at the output of the current controller (λ), the switches that must be turned ON and OFF, associated to phase A, are given by in Table 1. In this table, several possible conditions are presented, being the negative signal of the current the representation of the inversion of the current excitation in the winding.

Another aspect that must be considered is the control of the capacitor voltage. The capacitor voltage V_{C1} is controlled by an outer loop that gives the current reference. For this loop, a PI controller is adopted. The scheme of this outer loop can be seen in Figure 8.

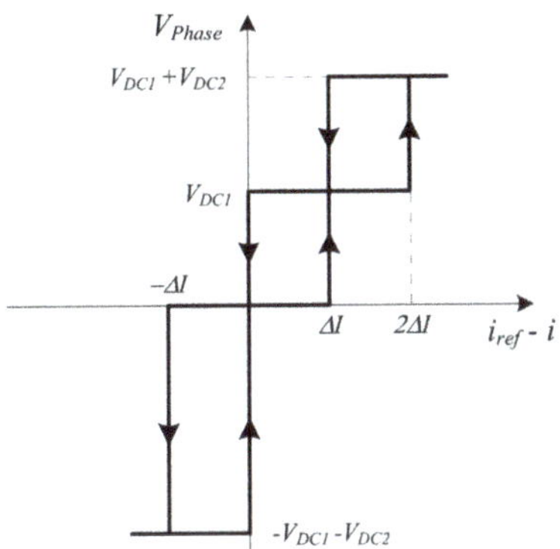

Figure 7. Multilevel hysteresis comparator associated with the current controller.

Table 1. Condition of the switches associated to phase A function of the output of the current controller.

λ	T_{1U}	T_{2U}	T_{1L}	T_{2L}	T_{3L}	T_{4L}	T_{5L}	T_{6L}	S_B	Current Direction	Voltage Level
4	1	0	1	0	0	0	1	0	0	+	$+V_{C1} + V_{C2}$
4	1	0	0	1	0	1	0	0	0	−	$+V_{C1} + V_{C2}$
4	0	1	1	0	0	0	1	0	1	+	$+V_{C1} + V_{C2}$
4	0	1	0	1	0	1	0	0	1	−	$+V_{C1} + V_{C2}$
3	0	0	1	0	0	0	1	0	0	+	$+V_{C1}$
3	0	0	0	1	0	1	0	0	0	−	$+V_{C1}$
2	0	0	1	0	0	0	0	0	0	+	0
2	0	0	0	0	0	0	1	0	0	+	0
2	0	0	0	1	0	0	0	0	0	−	0
2	0	0	0	0	0	1	0	0	0	−	0
1	0	0	0	0	0	0	0	0	0	+	$-V_{C1} - V_{C2}$
1	0	0	0	0	0	0	0	0	0	−	$-V_{C1} - V_{C2}$

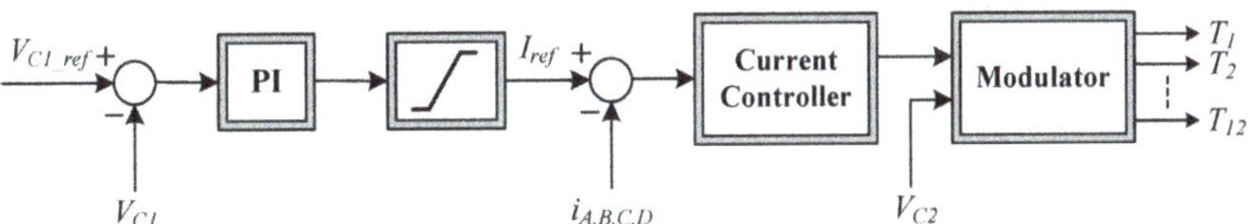

Figure 8. Control of the voltage capacitor V_{C1}.

Another aspect that needs to be considered is the balance of the floating capacitor voltage V_{C2}. This balance is ensured by the definition of a voltage reference for that capacitor and control of the switches T_{1U} and T_{2U} as a function of this reference. In this way, the gating signals of these switches are inhibited when the voltage of the capacitor C_2 is higher than the reference value.

5. Reliability Evaluation

This section presents a brief study about the theoretical probability of failure expected from the classic NPC-AHB [38] and the proposed fault-tolerant topology presented in Figure 2, after a failure in the IGBTs of one phase (winding A). In this analysis, the open-circuit failure mode and the ability to reach the desired voltage level were considered. Reliability measures the probability of a system failing within a given time interval $(0, t)$, i.e., reliability is a function of time, *R(t)* [49]. With a constant failure rate component, λ, reliability is determined by an exponential distribution (8):

$$R(t) = e^{-\lambda t} \tag{8}$$

The reliability of stand-by systems with redundancy, featuring ideal switching without repair and constant and identical failure rates, can be described by the Poisson distribution (9) [50].

$$R(t) = \sum_{k=0}^{n-1} \frac{(\lambda t)^2}{k!} e^{-\lambda t} \tag{9}$$

The probability of failure is then calculated by (10):

$$Q(t) = 1 - R(t) \tag{10}$$

Tables 2 and 3, below, present the estimated probability of failure after five years ($5 \times 8760 = 43800$ h) of operation and with a constant failure rate of $\lambda = 1.6577 \times 10^{-6}$ h^{-1} (failure rate according to Coffin–Manson method [51]) for the NPC-AHB and the proposed topology after the first failure regarding different power devices and desired voltage levels.

Table 2. Estimated probability of failure within 5 years after the first failure (NPC-AHB).

Device Failure	Open-Circuit Failures		
	+2 V	+V	0 V
T1A	Fail [1]	0.1957	0.070
T2A	Fail [1]	Fail [1]	0.1351
T3A	Fail [1]	Fail [1]	0.1351
T4A	Fail [1]	0.1957	0.070

[1] Fail—The converter fails immediately to achieve the desired voltage when the failure occurs in the specified power device.

Table 3. Estimated probability of failure within 5 years after the first failure considering the proposed topology.

Device Failure	Open-Circuit Failures		
	+2 V	+V	0 V
T1U	1.82×10^{-2}	2.33×10^{-5}	2.40×10^{-5}
T1L	4.93×10^{-2}	0.1351	3.43×10^{-4}
T5L	4.93×10^{-2}	0.1351	3.43×10^{-4}
T2U	9.46×10^{-3}	1.82×10^{-2}	2.40×10^{-5}
SB	9.46×10^{-3}	1.82×10^{-2}	2.40×10^{-5}

From this analysis, it is possible to conclude that the proposed solution is much more reliable regarding the open circuit failure mode than the classic NPC-AHB, independently of the device under failure and desired voltage level.

6. Simulation Results

The proposed PV generator feeding a water pumping system based on an SRM with a multilevel fault-tolerant converter was first tested by computer simulations. The simulation of this test system was performed through the use of MATLAB 2017a with Simulink version 8.9. Associated with this program, the Simulink Library, namely the Simscape/PowerSystems blockset, was also used. A discrete simulation type with a sampling time of 1 microsecond was also used. The model of the SRM was the model present in the examples of the Simulink. For the power semiconductors, the models present in the simscape/power systems/specialized technology/fundamental blocks/power electronics library were used. For these computer simulation tests, the capacitor C_1 was connected to a

PV generator that consisted of a PV panel and a classical DC–DC Boost converter. The PV panel allows the generation of a maximum power of 320 W. For the Boost and the proposed multilevel converter, an inductor of 0.4 mH and capacitors of 470 µF were considered.

The pumping system with the proposed multilevel power converter was initially tested under normal conditions. An irradiance of 1000 W/m^2 and a temperature of 25 °C were considered. For this test, a torque of 2.5 Nm and a speed of around 800 rpm were considered. The results of this test can be seen in Figure 9. This figure presents the voltage applied to the motor winding A, as well as the motor winding currents. Analyzing Figure 9a, it is possible to confirm the multilevel operation of the converter. Initially, when the motor winding A was excited, the maximum voltage was applied. Regarding the motor winding currents, it is possible to observe in Figure 9b that they were controlled and the reference value was around 4 A. Another aspect that was analyzed was the efficiency of the power electronic converter. Taking into consideration this operating mode, the efficiency was 92.8%. Regarding the power losses of the transistors, it is also possible to see that the lower transistors (T_{1L}, T_{2L}, T_{5L}, T_{6L}, T_{7U}, T_{8L}, T_{11L} and T_{12L}) were in the ON condition for much more time than the upper transistors (T_{1U} and T_{2U}). However, each of the upper transistors was common to two phases. Therefore, in this mode of operation, the power losses of the upper transistors were similar to the losses in the lower transistors. For the upper transistors, the power losses were around 46.2% of the total losses, while for the lower transistors, they were 53.8%.

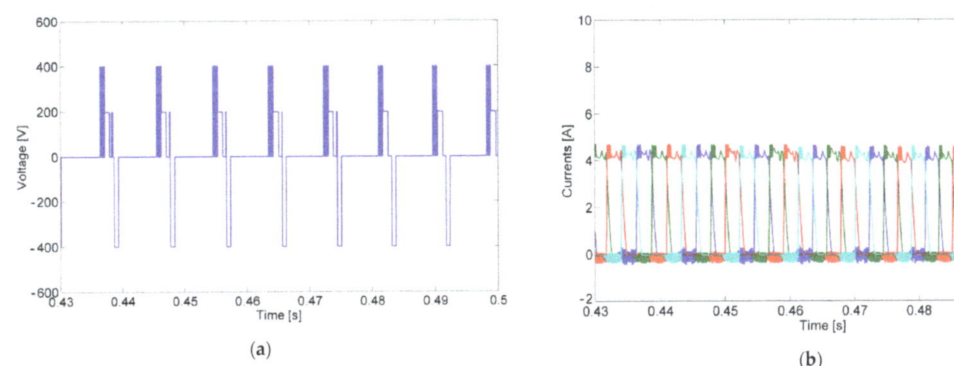

Figure 9. Simulation test with the converter operating in normal mode performed for an irradiance of 1000 W/m^2 and with a temperature of 25 °C: (**a**) winding voltage of phase A; (**b**) winding currents.

Another test in which the system was initially disconnected (capacitors discharged) and that turned ON when $t = 0$ s under the same irradiance and temperature conditions as the previous test, was performed to analyze the behavior of the proposed pumping system with the proposed converter and controllers. The result of the voltage across both capacitors can be seen in Figure 10. Through this figure, it is possible to confirm that initially, the capacitors were discharged and that, after connecting the system, they changed until they achieved the reference value. It is also possible to confirm that although, during the transient time, the voltages across the capacitors were different, they both stabilized with the same value (reference). In this way, it is possible to confirm that the controller ensured the balance between the capacitors.

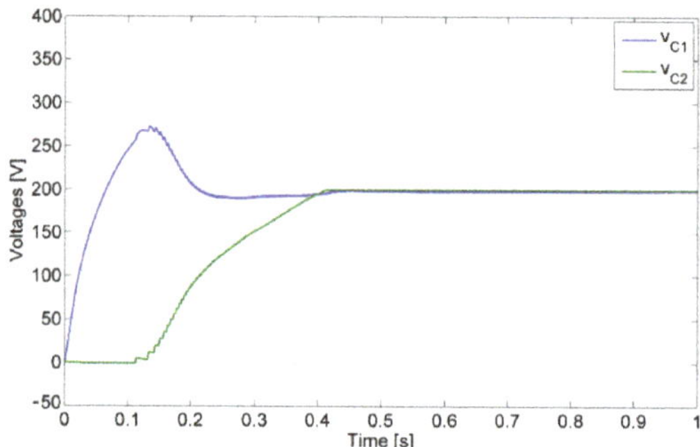

Figure 10. Voltages across the capacitors C_1 and C_2 during the connection period.

Tests with the proposed converter in healthy, faulty, and fault-tolerant conditions were also performed. A test in which, initially, the converter operated in healthy mode, following after $t = 0.5$ s with a fault in transistor T_{1U} and after $t = 0.52$ s in fault-tolerant mode, can be seen in Figure 11. This figure shows that after the switch fault, it was no longer possible to apply the maximum voltage ($V_{C1} + V_{C2}$). However, this limitation disappeared after the reconfiguration of the circuit (fault-tolerant mode). In fact, analyzing the winding currents, it is possible to verify that between 0.5 s and 0.52 s (during the fault) they were affected and began to decrease their values. However, after the reconfiguration of the circuit, the winding currents started to recover. Another test, with a different fault, was also performed. In this case, the fault was in one of the switches of the three-phase modules, namely transistor T_{1L}. The consequences of this fault and operation in fault-tolerant mode can be seen in Figure 12. After the fault ($t = 0.5$ s), it was not possible to apply a voltage to the SRM winding of phase A. Therefore, the current in this phase was zero. To overcome this problem, the fault-tolerant mode was applied at $t = 0.52$ s. In this last mode, the voltage was applied again to the SRM winding of phase A. However, as shown by the results of this last test, for this transistor fault, inverse voltages and negative currents to the SRM windings began to apply.

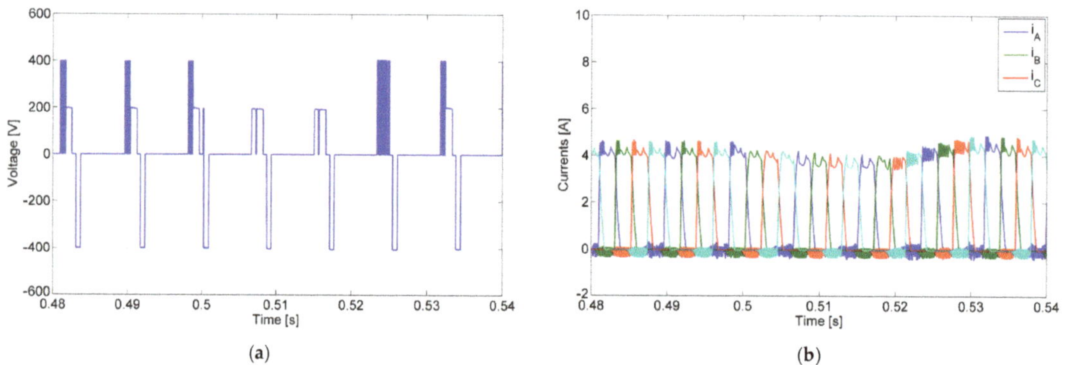

Figure 11. Simulation test with the converter operating in normal mode, fault in transistor T_{1U}, and fault-tolerant mode: (**a**) winding voltage of phase A; (**b**) winding currents.

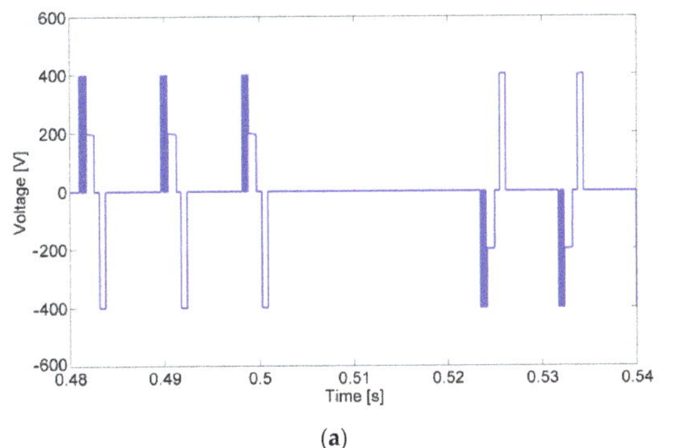 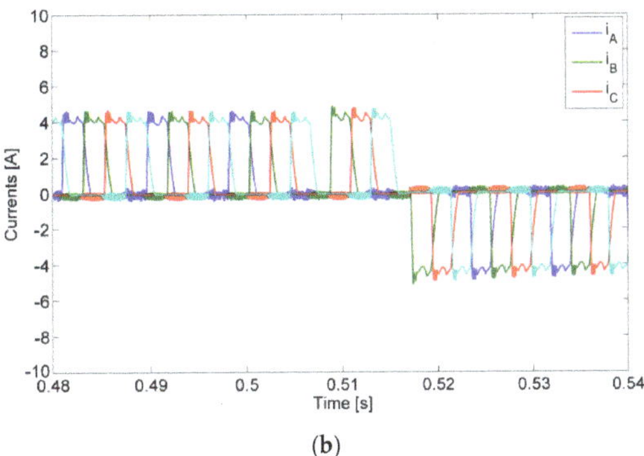

(**a**) (**b**)

Figure 12. Simulation test with the converter operating in normal mode, fault in transistor T_{1L}, and fault tolerant mode: (**a**) winding voltage of phase A; (**b**) winding currents.

7. Experimental Results

The verification of the simulation tests and theoretical considerations was also confirmed by some experimental tests using a laboratorial prototype. For the Boost and proposed multilevel converter similar components used in the simulation tests, an inductor of 0.4 mH and two capacitors of 470 µF were considered. To provide the DC voltage to the system, a controlled EA PS8360-30 2U (up to 360 V_{DC}, 30 A) power supply, adjusted to 200 V_{DC}, was used. This power supply makes it possible to emulate the PV generator (PV panels + DC–DC converter) in different weather conditions. The system was connected to a four-phase 8/6 SRM, the new proposed inverter, circuit drives, and sensors. The pumping system was emulated through a DC machine operating as a generator connected to the shaft of the four-phase 8/6 SRM. The control algorithm of the SRM drive was performed on a DSPACE tool. For the power semiconductors, IXGN72N60C3H1 and CM600DY-13T IGBTs and MUR1540 diodes were selected. The waveform signals were acquired by a TDS3014C oscilloscope.

The first experimental test was performed in normal conditions (without failures) considering a stationary voltage and current from the power supply. In this situation, the PV generator (PV panels + DC–DC converter) presented 200 V_{DC}, which corresponds to an irradiance of 1000 W/m^2, and the PV panels exhibited a temperature of 25 °C, boosted by the DC–DC converter. The first experimental result can be seen in Figure 13, showing the multilevel voltage applied to the motor winding A (Figure 13a) and the SRM winding currents (Figure 13b). Analyzing Figure 13, it is possible to confirm the desired multilevel operation of the converter and the controlled winding currents around the reference value (around 6 A). In this case, the current was slightly higher than the current presented in the simulations. However, this increase was due to the fact that the power supplied to the motor was adjusted to ensure the same speed, of 800 rpm. The efficiency of the power electronic converter in this operating mode was around 91.2%. This value was lower than the value obtained for the simulation, but it can be considered similar. Regarding the power losses of the transistors, they were also similar to the those obtained by simulation. In this case, the power losses of the upper transistors were around 43.8% of the total losses, while for the lower transistors, they were 56.2%.

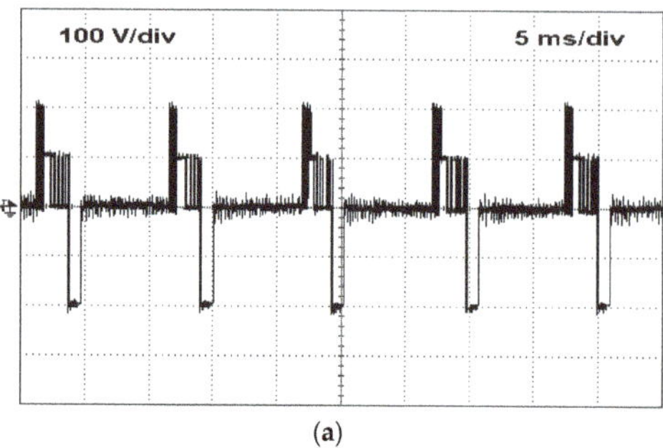
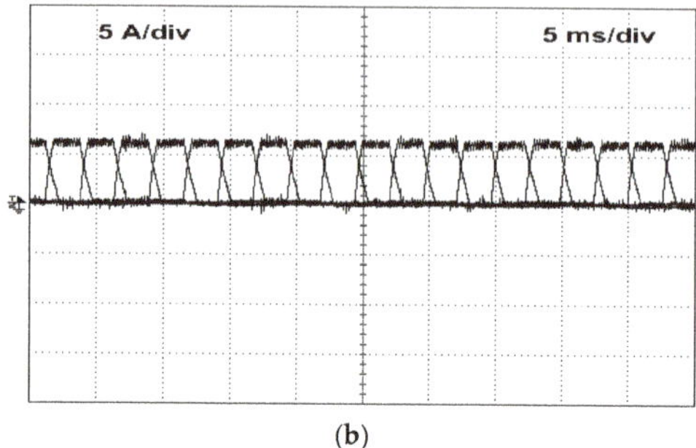

(**a**) (**b**)

Figure 13. Experimental results with the converter operating in normal mode (without failures) considering a stable voltage from the PV generator (200 V_{DC}): (**a**) multilevel voltage of winding phase A; (**b**) winding currents of the four-phase 8/6 SRM.

Another experimental test was performed to analyze the behavior of the proposed system considering the initial conditions. In this experimental test, the power supply was initially disconnected and then switched on to evaluate the transient voltage in the input capacitors of the multilevel converter. The experimental results of the voltage across both capacitors considering these initial conditions can be seen in Figure 14. Analyzing this figure, it is possible to observe that, initially, the capacitors were fully discharged and after switching ON the power source, the transient voltage increased until achieving the reference value. It can be seen that the adopted control strategy was able to stabilize and balance the voltages of both capacitors. The duration of this transient depended on the dynamics of several components and imposed load.

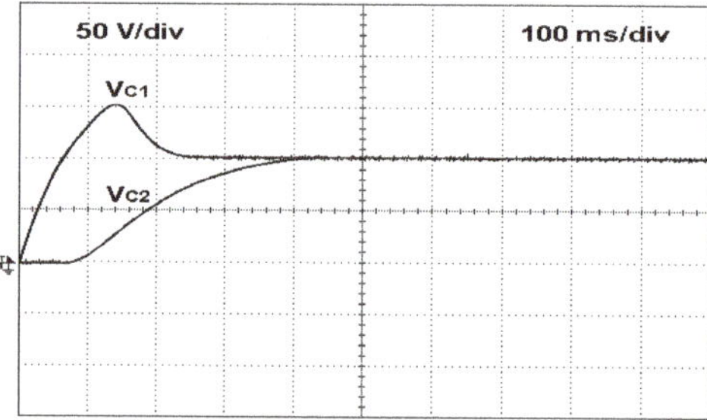

Figure 14. Experimental results of the voltages across the capacitors C_1 and C_2 during the initial transient after connecting the PV generator.

Another set of experimental tests was dedicated to analyzing the behavior of the system regarding the introduction of open-circuit failures in the power devices of the converter, making it possible to explore the features of the fault-tolerant operation. The first experimental result regarding the fault-tolerant operation can be seen in Figure 15. In this test, an open-circuit failure in the transistor T_{1U} at a certain moment in time was introduced, showing the change from healthy to fault operation during a couple of periods. After the fault detection and circuit reconfiguration, the converter started the fault-tolerant operation. Notably, that the open-circuit failure in the transistor T_{1U} generated the loss of the maximum voltage level ($+VDC$), since this device is an outer device connected to this voltage level. As a consequence, as demonstrated by the experimental results in Figure 15a, after the open-circuit failure of this device, the converter only operated with $+VDC/2$, which created difficulties for controlling the winding currents (Figure 15b). After the introduction of the fault-tolerant operation, the voltages and currents returned to normal operation. Notably, this failure mode did not require the inversion of the winding currents.

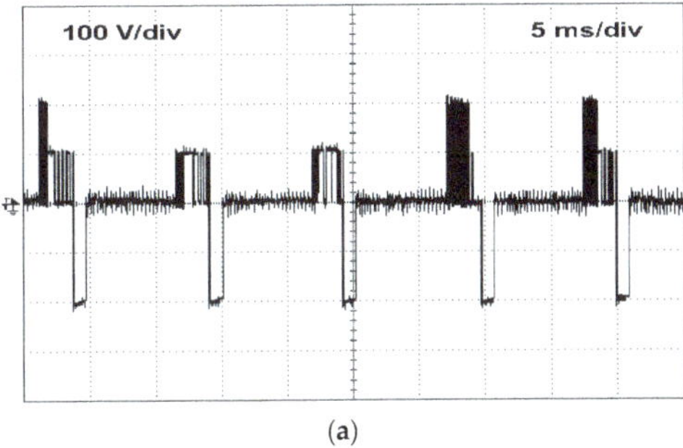
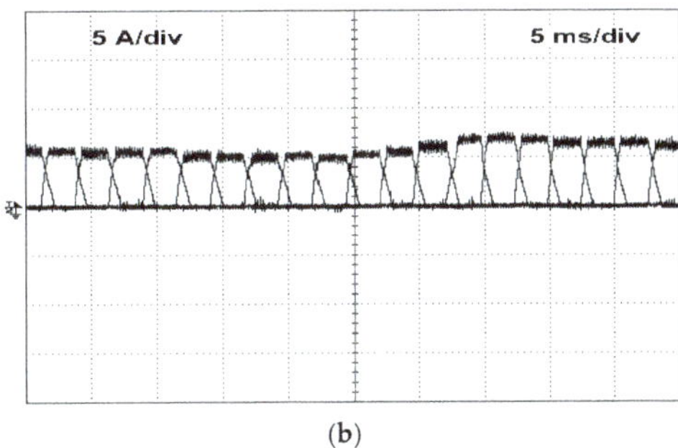

(a)   (b)

Figure 15. Experimental results with the converter operating in normal mode, followed by an open-circuit failure in the transistor the T_{1U} and finally fault-tolerant mode: (**a**) winding voltage of phase A; (**b**) winding currents.

A second experimental test of this kind was performed considering an open-circuit failure in one of the switches of the three-phase modules, the transistor T_{1L}. The experimental results of this test can be seen in Figure 16. In Figure 16a, it is possible to observe that this faut was introduced when the converter operated normally, creating a complete voltage loss in the winding of phase A. As a consequence of this fault, the voltage and current (Figure 16b) of phase A are missing. After fault detection and reconfiguration of the circuit, it was possible to observe that the multilevel voltages and current returned to the reference values. Nevertheless, the fault in this device (and other devices in the three-phase modules) required winding voltage and current inversion. This was possible to perform thanks to the design adopted for the topology, allowing the creation of alternative paths to recover voltage levels regardless of the polarity, exploiting the characteristics of the SRM to achieve the desired fault tolerance.

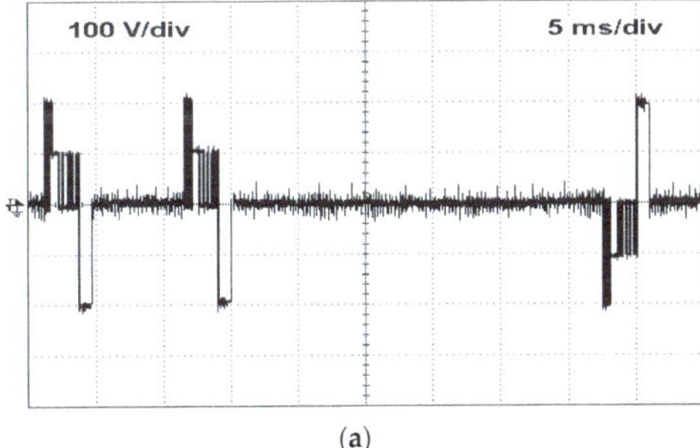
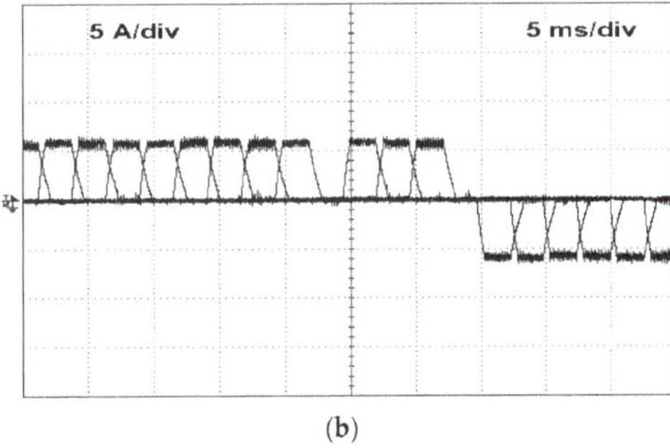

(a)   (b)

Figure 16. Experimental results with the converter operating in normal mode, followed by an open-circuit failure in the transistor the T_{1L} and, finally, fault-tolerant mode: (**a**) winding voltage of phase A; (**b**) winding currents.

Comparing these experimental results with those obtained by simulation, it is possible to verify that they are not precisely equal. However, it can be concluded that they are very similar. The small differences were expected, since in the simulation model, several components, such as the SRM, the power semiconductors, and the passive components, were simplified compared with the real components.

8. Discussion

As described above, a topology was designed to provide fault-tolerant capability but with a reduction of the power semiconductors and transistor drives in order to reduce the cost of the system. In Table 4, a comparison with equivalent topologies with fault-tolerant capability is presented. In fact, this new topology only requires sixteen switches and two extra diodes. The topology presented in [48] requires twenty-six switches and eight extra diodes. Furthermore, this topology also requires eight additional relays. Due to the number of switches and the requirement of additional relays, the cost of this converter is high. The topology presented in [49] requires 32 switches and 16 extra diodes. It does not require extra relays, but the number of switches is very high; accordingly, the cost is also high. A topology that also presents multilevel characteristics and a reduced number of switches is presented in [35]. This topology requires the same number of switches as the proposed topology, but a higher number of diodes. On the other hand, it does not provide full fault-tolerant operation, since for some faults, it does not offer the ability to restore the initial operation. Another aspect is the modularity, for which the topologies [35,47] present low abilities. On the other hand, the topology of [48] presents a high capability regarding the modularity, since it uses NPC legs that are also used in the drives. Regarding the proposed topology, it can be considered as medium, since in part of the circuit, it is possible to use classical three-phase two-level inverter modules.

Table 4. Comparison with equivalent multilevel topologies with fault-tolerant capability.

	Topology			
Characteristics	**[47]**	**[48]**	**[35]**	**Proposed**
Number of switches	32	26	16	16
Extra diodes	16	8	8	2
Additional relays	Yes	No	No	No
Converter modularity	Low	High	Low	Medium
Fault tolerant ability	Full	Full	Partial	Full
Cost	High	Medium	Low	Low

9. Conclusions

This work presented a PV generator that feeds a water pumping system based on an SRM. Associated with the SRM, a new multilevel converter is used. The topology was developed with the purpose of providing five voltage levels to the SRM windings. Besides the multilevel operation, this converter is also characterized by its ability to provide fault-tolerant operation during an open-switch fault. The converter was also developed with the purpose of reducing the number of power semiconductors and avoiding relays. To ensure this, the concept that in fault-tolerant mode, the change of the current excitation is also used was considered. On the other hand, the converter was based on two classical three-phase voltage source inverters, in which the four windings of the SRM are connected. All these aspects make it possible to reduce the cost of the fault-tolerant power converter for the SRM drive. The converter was also compared with other solutions. Furthermore, a study about the theoretical probability of failure expected from the SRM drive was also performed. The operation of the proposed converter in healthy and faulty mode was also analyzed. In addition, a control system and modulator for this converter were developed and presented. This system also ensures that the voltage across the two capacitors that is part of the converter is balanced. This paper also presents several test results performed with the proposed PV generator that feeds a water pumping system based on an SRM. The results confirm the ability of the proposed converter to operate with fault tolerance. These results also show the ability of the converter to control the currents in the SRM windings in both directions. This aspect proved to be essential in reducing the number of converter switches.

Author Contributions: Conceptualization, V.F.P. and A.J.P.; methodology, V.F.P.; validation, A.C., M.C. and D.F.; formal analysis, V.F.P., A.C. and A.J.P.; investigation, V.F.P. and A.C.; writing—original draft preparation, V.F.P. and A.J.P.; writing—review and editing, D.F., M.C. and A.J.P.; visualization, D.F., A.J.P. and M.C.; supervision, A.J.P. All authors have read and agreed to the published version of the manuscript.

Funding: This research was funded by national funds through FCT-Fundação para a Ciência e a Tecnologia, under projects UIDB/50021/2020 and UIDB/00066/2020.

Institutional Review Board Statement: Not applicable.

Informed Consent Statement: Not applicable.

Data Availability Statement: Not applicable.

Acknowledgments: This research was supported by national funds through FCT-Fundação para a Ciência e a Tecnologia, under projects UIDB/50021/2020 and UIDB/00066/2020.

Conflicts of Interest: The authors declare no conflict of interest.

References

1. Periasamy, P.; Jain, N.; Singh, I. A review on development of photovoltaic water pumping system. *Renew. Sustain. Energy Rev.* **2015**, *43*, 918–925. [CrossRef]
2. Angadi, S.; Yaragatti, U.R.; Suresh, Y.; Raju, A.B. Comprehensive review on solar, wind and hybrid wind-PV water pumping systems-an electrical engineering perspective. *CPSS Trans. Power Electron. Appl.* **2021**, *6*, 1–19. [CrossRef]
3. Stoyanov, L.; Bachev, I.; Zarkov, Z.; Lazarov, V.; Notton, G. Multivariate Analysis of a Wind–PV-Based Water Pumping Hybrid System for Irrigation Purposes. *Energies* **2021**, *14*, 3231. [CrossRef]
4. Muralidhar, K.; Rajasekar, N. A review of various components of solar water-pumping system: Configuration, characteristics, and performance. *Int. Trans. Electr. Energy Syst.* **2021**, *31*, 1–33. [CrossRef]
5. Shepovalova, O.V.; Belenov, A.T.; Chirkov, S.V. Review of photovoltaic water pumping system research. *Energy Rep.* **2020**, *6*, 306–324. [CrossRef]
6. Chandel, S.; Naik, M.N.; Chandel, R. Review of solar photovoltaic water pumping system technology for irrigation and community drinking water supplies. *Renew. Sustain. Energy Rev.* **2015**, *49*, 1084–1099. [CrossRef]
7. El-Kharashi, E. Design and predicting efficiency of highly nonlinear hollow cylinders switched reluctance motor. *Energy Convers. Manag.* **2007**, *48*, 2261–2275. [CrossRef]
8. Narayana, V.; Mishra, A.K.; Singh, B. Development of low-cost PV array-fed SRM drive-based water pumping system utilizing CSC converter. *IET Power Electron.* **2017**, *10*, 156–168. [CrossRef]
9. Cordeiro, A.; Fernão Pires, V.; Foito, D.; Pires, A.J.; Martins, J.F. Three-level quadratic boost DC-DC converter associated to a SRM drive for water pumping photovoltaic powered systems. *Sol. Energy* **2020**, *209*, 42–56. [CrossRef]
10. Han, G.; Chen, H.; Guan, G. Low-cost SRM drive system with reduced current sensors and position sensors. *IET Electr. Power Appl.* **2019**, *13*, 853–862. [CrossRef]
11. Mishra, A.K.; Singh, B. Efficient solar-powered water pump with single-input dual-output DC–DC converter employing four-phase SRM drive. *IET Power Electron.* **2020**, *13*, 3435–3444. [CrossRef]
12. Koreboina, V.B.; Narasimharaju, B.L.; Vinod Kumar, D.M. Performance evaluation of switched reluctance motor PWM control in pv-fed water pump system. *Int. J. Renew. Energy Res.* **2016**, *6*, 941–950.
13. Hrabovcová, V.; Rafajdus, P.; Lipták, M.; Szabó, L. Performance of converters suitable for switched reluctance generator (SGR) operation. *J. Electr. Eng.* **2013**, *64*, 201–211.
14. Pires, V.F.; Pires, A.J.; Cordeiro, A.; Foito, D. A Review of the Power Converter Interfaces for Switched Reluctance Machines. *Energies* **2020**, *13*, 3490. [CrossRef]
15. Ellabban, O.; Abu-Rub, H. Switched reluctance motor converter topologies: A review. In Proceedings of the IEEE International Conference on Industrial Technology, Busan, Korea, 26 February–1 March 2014; pp. 840–846.
16. Barnes, M.; Pollock, C. Power electronic converters for switched reluctance drives. *IEEE Trans. Power Electron.* **1998**, *13*, 1100–1111. [CrossRef]
17. Zhang, Z.; Cheung, N.C.; Cheng, K.W.E.; Xue, X.D.; Lin, J.K.; Bao, Y.J. Analysis and design of a cost effective converter for switched reluctance motor drives using component sharing. In Proceedings of the 2011 4th International Conference on Power Electronics Systems and Applications, Hong Kong, China, 8–10 June 2011; pp. 1–6.
18. Tomczewski, K.; Wrobel, K. Improved C-dump converter for switched reluctance motor drives. *IET Power Electron.* **2014**, *7*, 2628–2635. [CrossRef]
19. Dhumal, K.R.; Dhamse, S.S. Solar PV array based water pumping by using SRM drive: A review. In Proceedings of the 2018 International Conference on Computation of Power, Energy, Information and Communication (ICCPEIC), Chennai, India, 28–29 March 2018; pp. 140–146.

20. Kobler, R.; Andessner, D.; Amrhein, W. Development of a SRM power electronic system with a reduced number of power semiconductors. In Proceedings of the 2011 3rd International Youth Conference on Energetics (IYCE), Leiria, Portugal, 7–9 July 2011; pp. 1–7.

21. Gan, C.; Chen, Y.; Qu, R.; Yu, Z.; Kong, W.; Hu, Y. An Overview of Fault-Diagnosis and Fault-Tolerance Techniques for Switched Reluctance Machine Systems. *IEEE Access* **2019**, *7*, 174822–174838. [CrossRef]

22. Cordeiro, A.; Pires, V.F.; Pires, A.J.; Martins, J.F.; Chen, H. Fault-tolerant voltage-source-inverters for switched reluctance motor drives. In Proceedings of the 2019 IEEE 13th International Conference on Compatibility, Power Electronics and Power Engineering (CPE-POWERENG), Sonderborg, Denmark, 23–25 April 2019; pp. 1–6.

23. Sun, Q.; Wu, J.; Gan, C.; Guo, J. Modular full-bridge converter for three-phase switched reluctance motors with integrated fault-tolerance capability. *IEEE Trans. Power Electron.* **2019**, *34*, 2622–2634. [CrossRef]

24. Fang, C.; Chen, H. Design rule for fault-tolerant converters of switched reluctance motors. *J. Power Electron.* **2021**, *21*, 1690–1700. [CrossRef]

25. Chen, Q.; Xu, D.; Xu, L.; Wang, J.; Lin, Z.; Zhu, X. Fault-tolerant operation of a novel dual-channel switched reluctance motor using two 3-phase standard inverters. *IEEE Trans. Appl. Supercond.* **2018**, *28*, 1–5. [CrossRef]

26. Mohamed, A.H.; Vansompel, H.; Sergeant, P. Reconfigurable Modular Fault-Tolerant Converter Topology for Switched Reluctance Motors. *IEEE J. Emerg. Sel. Top. Power Electron.* **2021**. [CrossRef]

27. Azer, P.; Ye, J.; Emadi, A. Advanced Fault-Tolerant Control Strategy for Switched Reluctance Motor Drives. In Proceedings of the 2018 IEEE Transportation Electrification Conference and Expo (ITEC), Long Beach, CA, USA, 20–25 June 2018.

28. Ali, N.; Gao, Q.; Xu, C.; Makys, P.; Stulrajter, M. Fault diagnosis and tolerant control for power converter in SRM drives. *J. Eng.* **2018**, *13*, 546–551. [CrossRef]

29. Hu, Y.; Gan, C.; Cao, W.; Li, W.; Finney, S.J. Central-Tapped Node Linked Modular Fault-Tolerance Topology for SRM Applications. *IEEE Trans. Power Electron.* **2016**, *31*, 1541–1554. [CrossRef]

30. Hu, Y.; Gan, C.; Cao, W.; Zhang, J.; Li, W.; Finney, S.J. Flexible Fault-Tolerant Topology for Switched Reluctance Motor Drives. *IEEE Trans. Power Electron.* **2016**, *31*, 4654–4668. [CrossRef]

31. Ding, W.; Hu, Y.; Wu, L. Investigation and Experimental Test of Fault-Tolerant Operation of a Mutually Coupled Dual Three-Phase SRM Drive Under Faulty Conditions. *IEEE Trans. Power Electron.* **2015**, *30*, 6857–6872. [CrossRef]

32. Gameiro, N.S.; Marques Cardoso, A.J. Fault tolerant power converter for switched reluctance drives. In Proceedings of the 2008 18th International Conference on Electrical Machines, Vilamoura, Portugal, 6–9 September 2008; pp. 1–6.

33. Pires, V.F.; Amaral, T.G.; Cordeiro, A.; Foito, D.; Pires, A.J.; Martins, J.F. Fault-tolerant SRM drive with a diagnosis method based on the entropy feature approach. *Appl. Sci.* **2020**, *10*, 3516. [CrossRef]

34. Patil, D.; Wang, S.; Gu, L. Multilevel converter topologies for high-power high-speed switched reluctance motor: Performance comparison. In Proceedings of the 2016 IEEE Applied Power Electronics Conference and Exposition (APEC), Long Beach, CA, USA, 20–24 March 2016; pp. 2889–2896.

35. Ma, M.; Yuan, K.; Yang, Q.; Yang, S. Open-circuit fault-tolerant control strategy based on five-level power converter for SRM system. *CES Trans. Electr. Mach. Syst.* **2019**, *3*, 178–186. [CrossRef]

36. Azer, P.; Bauman, J. An Asymmetric Three-Level T-Type Converter for Switched Reluctance Motor Drives in Hybrid Electric Vehicles. In Proceedings of the IEEE Transportation Electrification Conference and Expo, Novi, MI, USA, 19–21 June 2019; pp. 1–6.

37. Scholtz, P.A.; Gitau, M.N. Carrier Modulation Schemes of Asymmetric, Multileveled, Switched Reluctance Machine Drives. In Proceedings of the 22nd IEEE International Conference on Industrial Technology (ICIT), Valencia, Spain. 10–12 March 2021; pp. 160–165.

38. Peng, F.; Ye, J.; Emadi, A. An Asymmetric Three-Level Neutral Point Diode Clamped Converter for Switched Reluctance Motor Drives. *IEEE Trans. Power Electron.* **2017**, *32*, 8618–8631. [CrossRef]

39. Gan, C.; Sun, Q.; Wu, J.; Kong, W.; Shi, C.; Hu, Y. MMC-Based SRM Drives With Decentralized Battery Energy Storage System for Hybrid Electric Vehicles. *IEEE Trans. Power Electron.* **2019**, *34*, 2608–2621. [CrossRef]

40. Abdel-Aziz, A.A.; Ahmed, K.H.; Wang, S.; Massoud, A.M.; Williams, B.W. A Neutral-Point Diode-Clamped Converter With Inherent Voltage-Boosting for a Four-Phase SRM Drive. *IEEE Trans. Ind. Electron.* **2020**, *67*, 5313–5324. [CrossRef]

41. Hu, Y.; Gan, C.; Cao, W.; Li, C.; Finney, S.J. Split Converter-Fed SRM Drive for Flexible Charging in EV/HEV Applications. *IEEE Trans. Ind. Electron.* **2015**, *62*, 6085–6095. [CrossRef]

42. Song, S.; Peng, C.; Guo, Z.; Ma, R.; Liu, W. Direct Instantaneous Torque Control of Switched Reluctance Machine Based on Modular Multi-Level Power Converter. In Proceedings of the 22nd International Conference on Electrical Machines and Systems, Harbin, China, 11–14 August 2019; pp. 1–6.

43. Pires, V.F.; Cordeiro, A.; Pires, A.J.; Martins, J.F.; Chen, H. A multilevel topology based on the T-type converter for SRM drives. In Proceedings of the 2018 16th Biennial Baltic Electronics Conference (BEC), Tallinn, Estonia, 8–10 October 2018; pp. 1–4.

44. Lee, D.; Wang, H.; Ahn, J. An advanced multi-level converter for four-phase SRM drive. In Proceedings of the 2008 IEEE Power Electronics Specialists Conference, Rhodes, Greece, 15–19 June 2008; pp. 2050–2056.

45. Tang, Y.; He, Y.; Wang, F.; Xie, H.; Rodríguez, J.; Kennel, R. A Drive Topology for High-Speed SRM With Bidirectional Energy Flow and Fast Demagnetization Voltage. *IEEE Trans. Ind. Electron.* **2021**, *68*, 9242–9253. [CrossRef]

46. Lee, D.; Ahn, J. A Novel Four-Level Converter and Instantaneous Switching Angle Detector for High Speed SRM Drive. *IEEE Trans. Power Electron.* **2007**, *22*, 2034–2041. [CrossRef]

47. Pires, V.F.; Cordeiro, A.; Foito, D.; Pires, A.J. Fault-Tolerant Multilevel Converter to Feed a Switched Reluctance Machine. *Machines* **2022**, *10*, 35. [CrossRef]
48. Pires, V.F.; Cordeiro, A.; Foito, D.; Pires, A.J.; Martins, J.; Chen, H. A multilevel fault-tolerant power converter for a switched reluctance machine drive. *IEEE Access* **2020**, *8*, 21917–21931. [CrossRef]
49. Rausand, M.; Hoyland, A. *System Reliability Theory—Models, Statistical Methods, and Applications*, 2nd ed.; John Wiley & Sons: Hoboken, NJ, USA, 2004; ISBN-13 978-0471471332.
50. Hardas, R.G.; Munshi, A.P.; Kadwane, S.G. Reliability of different levels of cascaded H-Bridge inverter: An investigation and comparison. In Proceedings of the IEEE International Conference on Advanced Communications, Control and Computing Technologies, Ramanathapuram, India, 8–10 May 2014; pp. 349–354.
51. Smith, D.J. *Reliability Mantainability and Risk—Practical Method for Engineers*, 6th ed.; Elsevier Butterworth-Heinemann: Burlington, MA, USA, 2001.

Review

Continuous Control Set Model Predictive Control of a Switch Reluctance Drive Using Lookup Tables

Alecksey Anuchin [1,2,*], Galina L. Demidova [2], Chen Hao [3], Alexandr Zharkov [1], Andrei Bogdanov [2] and Václav Šmídl [4]

[1] Department of Electric Drives, Moscow Power Engineering Institute, 111250 Moscow, Russia; Zharkov@mpei.ru
[2] Faculty of Control Systems and Robotics, ITMO University, 197101 Saint Petersburg, Russia; demidova@itmo.ru (G.L.D.); anbogdanov@itmo.ru (A.B.)
[3] School of Electrical and Power Engineering, China University of Mining and Technology, Xuzhou 221116, China; hchen@cumt.edu.cn
[4] Department of Adaptive Systems, Institute of Information Theory and Automation, CZ-182 00 Prague, Czech Republic; smidl@utia.cas.cz
* Correspondence: anuchin.alecksey@gmail.com

Received: 18 May 2020; Accepted: 22 June 2020; Published: 29 June 2020

Abstract: A problem of the switched reluctance drive is its natural torque pulsations, which are partially solved with finite control set model predictive control strategies. However, the continuous control set model predictive control, required for precise torque stabilization and predictable power converter behavior, needs sufficient computation resources, thus limiting its practical implementation. The proposed model predictive control strategy utilizes offline processing of the magnetization surface of the switched reluctance motor. This helps to obtain precalculated current references for each torque command and rotor angular position in the offline mode. In online mode, the model predictive control strategy implements the current commands using the magnetization surface for fast evaluation of the required voltage command for the power converter. The proposed strategy needs only two lookup table operations requiring very small computation time, making instant execution of the whole control system possible and thereby minimizing the control delay. The proposed solution was examined using a simulation model, which showed precise and rapid torque stabilization below rated speed.

Keywords: switched reluctance motor drive; model predictive control; continuous control set; pulse-width modulation; magnetization surface; electrical drive

---

1. Introduction

Switched reluctance drives (SRDs) have attracted a significant amount of attention in recent decades as the most promising type of electric drive. With the spread of 3D-printing technologies, this drive is one of the best candidates to become the first commercial 3D-printed machine [1] and modular designed machine [2]. Being very simple, the machine requires a sophisticated control strategy, which currently has no general approach, unlike the field-oriented control (FOC) or direct torque control (DTC) used in AC electric drives. Many papers have considered the mitigation of torque ripple by adjustment of the commutation angles [3–5] and current profiles [6,7], or optimization of the motor magnetic geometry [8]. However, solutions suggested in [3–5] help in the limited range of speeds as the current slope varies with the speed also affecting the produced torque. Current profiling suggested in [6,7] cannot be applied to each particular motor without hand tuning in all operation modes. In addition, optimization of the magnetic geometry for torque ripple minimization [8] contradicts the goal of efficiency optimization of the electrical machine.

The approach with model predictive control using a finite control set (FCS) was suggested in [9] to help decrease torque ripple and make the control strategy applicable for any motor. It was improved in [10] by decreasing the losses in the motor and in [11] by achieving equal distribution of the temperatures of the power converter's IGBT-modules. These approaches utilize the magnetization surface of the switched reluctance motor for torque estimation while selecting the best control command. These systems stabilize the output torque by profiling the phase current shape, not only with respect to the commanded torque, but also minimizing ohmic losses. The disadvantage of the finite control set model predictive control is the random switching rate and the error in the commanded torque or speed if a short prediction horizon is used. Continuous control set (CCS) model predictive control solves the voltage equations to obtain the references to the inverter, but this approach is problematic for a switched reluctance machine (SRM) due to the high nonlinearity of the controlled plant as the motor inductance varies with current and the power converter frequently saturates.

During analysis of the model predictive control (MPC) operation it was noticed that, for each electrical revolution for the same torque command, the current references repeat. The cost function, which considers the difference between the commanded and predicted torque, and minimizes ohmic losses, gives the predefined shape of the current references with respect to the rotor position. Together with the magnetization surface of each phase, this can be used to evaluate the desired voltage command, which adjusts the state variables of the drive to follow the references. Compared to the existing methods, the proposed solution uses information about the magnetization map of the machine to obtain the current references for the demanded torque, and then evaluates voltage references using the magnetization surface. The control strategy uses only lookup table operations, which allows evaluation of the voltage command nearly instantly, thus requiring very small computation efforts.

2. Switched Reluctance Drive Model

2.1. Drive Topology

The switched reluctance drive can have a varying number of phases and its power converter topology can vary. The most common configuration of the drive is shown in Figure 1. The stator of the motor contains three phases with concentrated windings located at six stator teeth. The rotor has four teeth, and the electromechanical reduction ratio is 4. The power converter is represented by three asymmetrical H-bridges, which feed the motor phases with unipolar current.

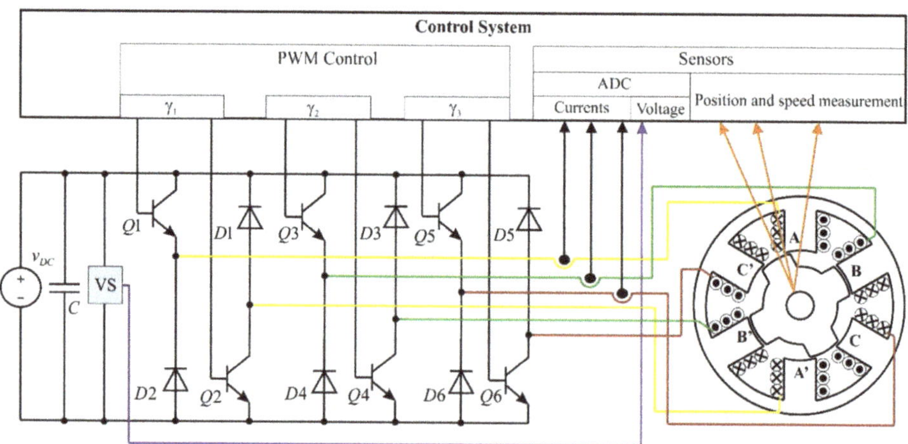

Figure 1. Switched reluctance drive topology.

2.2. Motor Equations

Each phase has its voltage balance equation [10]:

$$\frac{d\psi}{dt} = v - iR, \tag{1}$$

where v is the applied voltage, i is the flowing current, R is the phase resistance, and ψ is the flux linkage of the winding.

The magnetic system of an SRM is highly non-linear, which is usually defined by means of a magnetization surface/map or flux linkage map. For AC machines these maps can be identified online using a testbench that consists of the motor under test and a prime mover motor [12]. This testbench can realize all possible operation conditions and uses high-frequency injection to estimate differential inductance maps in order to build flux linkage maps. Such a method is applicable for SRMs; however, other methods are more popular due to the specific nature of this machine.

The motor phases of an SRM are usually considered to be independent without, or with very little, magnetic coupling between them [11]. Therefore, the magnetization surface of each phase depends on its flowing current and current rotor angular position as shown in Figure 2. The magnetization map of an SRM can be obtained experimentally [12,13]. Paper [12] provides a thorough description of experimental setups for making flux linkage maps. In [13], authors represent two methods of flux linkage map measurement based on the example of a four-phase SRM. The first method utilized a testbench with a mechanical rotor locking device and measured current and voltage. The flux linkage was calculated offline using experimental data. The second method was an online method. It did not require a rotor locking device, and online data from current and voltages sensors were used to evaluate the magnetization map when the SRM was running in a normal operation mode. For the considered solution, the offline method is preferable, as the online method provides an incomplete data set limited by the operation conditions of the drive, which makes estimation of the torque surface impossible.

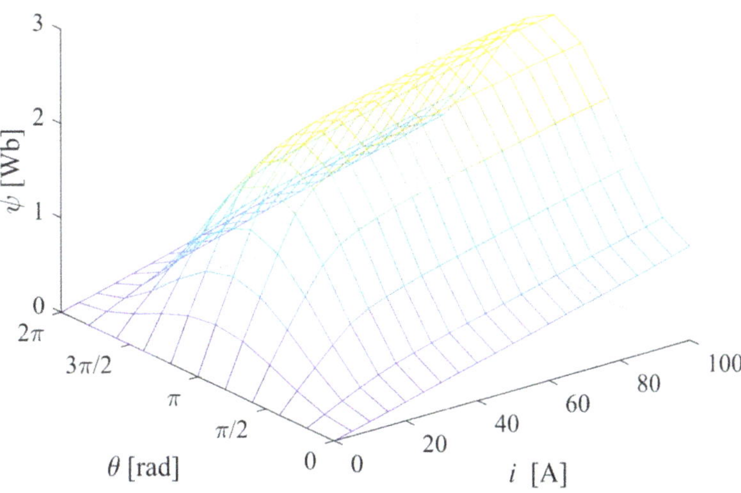

Figure 2. Magnetization surface of the switched reluctance motor (SRM).

The torque of the phase at the electrical speed can be defined as a function of the phase energy:

$$T = -\frac{\partial \int_0^\psi i(\psi, \theta)d\psi}{\partial \theta},$$ (2)

or co-energy:

$$T = \frac{\partial \int_0^i \psi(i, \theta)di}{\partial \theta},$$ (3)

where θ is the electrical rotor angular position.

The total torque of the machine is the sum of the torques produced by each phase. For a considered motor with number of phases N equal to 3, the torque equation can be written as follows:

$$T = \sum_{n=1}^{N} T_n = T_1 + T_2 + T_3.$$ (4)

2.3. Power Converter

Each phase of the motor is connected to the asymmetrical bridge as shown in Figure 1. The phase current can only flow in one direction, which is considered to be the positive direction. It is possible to apply a positive voltage of the DC link by switching on both switches in the bridge, zero voltage by switching on only one of the switches, and negative voltage by switching off both switches. In the last case the phase current flows through freewheeling diodes, and the negative voltage is applied only while the flowing current is bigger than zero.

3. Continuous Control Set Model Predictive Control for SRD

The FCS MPC, which is taken as the reference, was proposed in [10]. The MPC algorithm checked 27 possible states of the power converter and estimated the behaviors of the phase currents and torque for each case. The cost function was developed to keep total torque closer to the reference with minimum ohmic losses in the phase windings.

As a result, for the infinite switching rate, the current shape was profiled to achieve nearly constant torque as shown in Figure 3a. The waveform of the phase current contains the period of growing current, when the phase operates together with the previously on phase, which is reaching the point of maximum inductance (aligned position). When the previous phase reaches the aligned position, it is not desirable to keep current in it anymore as it does not produce positive torque. So, the previously on phase turns off, and all the torque, produced by the motor, is delivered to the shaft by a single phase at its second time interval. The current is regulated according to the magnetization surface of the motor to produce the commanded torque. Then, the inductance of this phase reaches its maximum value. Its current is reduced, and the next phase is switched on.

This current waveform is repeated each revolution; therefore, it is possible to evaluate it in advance (offline) for each torque reference.

Having variable rotor angular positions and torque references, the solution for each phase current can be represented as a function of these two variables. This function is easy to implement using a lookup table. An example of the current reference surface for phase A is shown in Figure 3b.

The current regulation can be performed using the magnetization surface of the motor phase. With the rotation of the machine and change of the current reference, the system should move from one point of the surface to another.

At the end of the pulse-width modulation (PWM) cycle, the current is to be measured and the control system should be executed. The control system has information about the measured current

$i[k]$ (see Figure 4) and current rotor electrical position $\theta[k]$. Knowing the motor speed ω, the rotor position in the end of the next PWM cycle can be evaluated by:

$$\hat{\theta}[k+1] = \theta[k] + \omega \cdot T_{PWM}, \tag{5}$$

where T_{PWM} is the duration of the PWM cycle.

Knowing the torque reference T_{ref}, the current reference for the next PWM cycle can be evaluated from the lookup table as:

$$i_{ref}[k+1] = f_{\text{current reference}}\left(\hat{\theta}[k+1], T_{ref}\right). \tag{6}$$

Thus, the control system knows the phase currents and current rotor position, and current references and predicted rotor position for the next PWM cycle. The system moves along the magnetization surface of the motor phase from point k to point $k+1$ as shown in Figure 4.

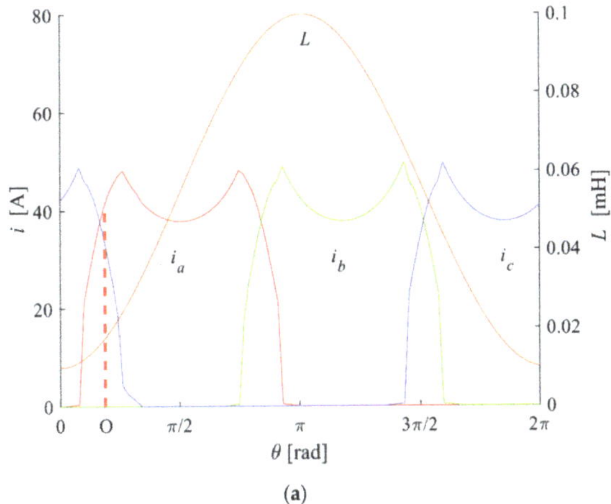

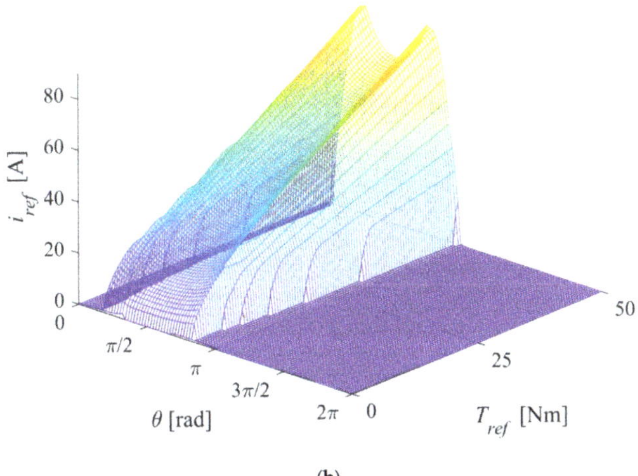

Figure 3. Current waveform for constant torque operation (**a**) and the phase A current reference surface example (**b**).

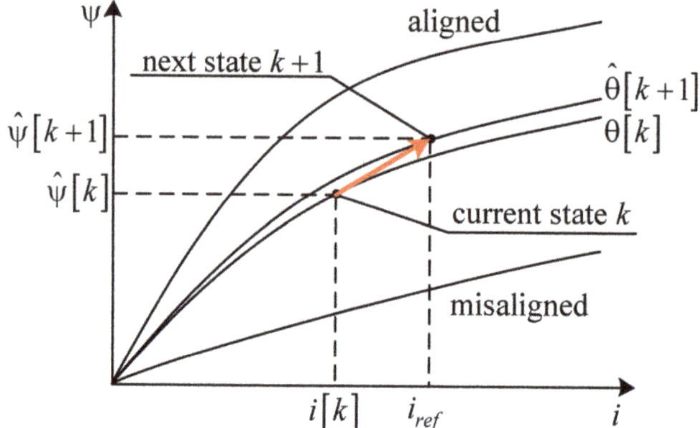

Figure 4. Transient during a single pulse-width modulation (PWM) cycle.

For both points the flux linkage can be estimated as:

$$\left.\begin{array}{l}\hat{\psi}[k] = f_{\text{magnetization surface}}(\theta[k], i[k]); \\ \hat{\psi}[k+1] = f_{\text{magnetization surface}}(\hat{\theta}[k+1], i_{ref}).\end{array}\right\} \tag{7}$$

These data can now be used to solve Equation (1) in order to find the voltage command:

$$v_{ref} = \frac{\hat{\psi}[k+1] - \hat{\psi}[k] - \frac{i[k]+i_{ref}}{2}R \cdot T_{PWM}}{T_{PWM}}, \tag{8}$$

which depends on the difference between the next and current flux linkage estimations taking into account the voltage drop across the phase resistance by the average current that will flow in the winding during the next PWM cycle.

The duty cycle for the asymmetrical bridge control can be evaluated using the voltage command and the current DC link voltage v_{DC}:

$$\gamma = \frac{v_{ref}}{v_{DC}}, \tag{9}$$

which should be limited in the range of $[-1; +1]$, where $+1$ corresponds to fully on switches and -1 corresponds to fully off.

4. Control System Implementation

4.1. Evaluation of the Magnetization Surface and Torque Surface as the Functions of Phase Current and Rotor Angular Position

The surface from Figure 2 should be represented by an array of $N \cdot N$ points. As the resulting array lies in the memory of the control system (microcontroller or field-programmable gate array (FPGA)), the number of points N should be selected taking into account this constraint as well as the accuracy of the surface representation. If the number of points is sufficiently large, it is possible to use simple bilinear interpolation or fetch the nearest value for fast evaluation of the control Equation (7). The array of 100 points in both dimensions requires 10,000 words of memory for surface representation, which is suitable for most modern microcontroller devices. This representation is convenient if the online estimation of the flux linkages is running in parallel with the control system adjusting the reference points in the magnetization map [14].

Having a magnetization surface representation, it is possible to obtain the torque surface, which can be also represented as an array of $N \cdot N$ points. The torque can be evaluated as the derivative of the co-energy with respect to the angle. The co-energy integral can be expressed with series for numerical evaluation using Tustin's method as:

$$\Delta W'\left(i_{N_i}, \theta_{N_\theta}\right) = \int_0^{i_{N_i}} \psi\left(i, \theta_{N_\theta}\right) di \approx \sum_{n_i=2}^{N_i} \left(\frac{\psi[n_i-1, N_\theta] + \psi[n_i, N_\theta]}{2} \right), \tag{10}$$

where i_{N_i} and θ_{N_θ} are the current and angle of N_i- and N_θ-points in the magnetization surface table, respectively. Each point on the surface can be evaluated as a difference between co-energies in a one-step clockwise direction from the current rotor angular position and a one-step counterclockwise direction divided by the angle difference using the following equation:

$$\begin{aligned}
T[N_i, N_\theta] &= \frac{\Delta W'(i,\theta)}{\Delta\theta} \\
&= \frac{\sum\limits_{n_i=2}^{N_i}\left(\left(\frac{\psi[n_i-1,N_\theta+1]+\psi[n_i,N_\theta+1]}{2}\right)\cdot\frac{I_{max}}{N}\right) - \sum\limits_{n_i=1}^{N_i}\left(\left(\frac{\psi[n_i-1,N_\theta-1]+\psi[n_i,N_\theta-1]}{2}\right)\cdot\frac{I_{max}}{N}\right)}{\left(2\cdot\frac{2\pi}{N}\right)},
\end{aligned} \tag{11}$$

which can be simplified as:

$$T[N_i, N_\theta] = \frac{I_{max}}{8\pi} \cdot \sum_{n_i=2}^{N_i} (\psi[n_i-1, N_\theta+1] + \psi[n_i, N_\theta+1] - \psi[n_i-1, N_\theta-1] - \psi[n_i, N_\theta-1]). \tag{12}$$

4.2. Evaluation of the Current Reference Surface

The referenced torque can be produced by an infinite number of combinations of the phase currents. If minimum ohmic losses are desired, then there is a single solution for a given torque reference and angular position. Thus, the goal is to evaluate the current reference for each phase as a function of the torque reference and the rotor angular position.

For the three-phase SRM, there are angular positions when only one phase can produce a positive torque, or positions when two phases can produce it. Consider the situation for two phases generating positive torque.

For the used linear model of the motor, it is possible to find an analytical solution for the optimal current references, but this model is used only to simplify the simulation model. In the general case, the magnetization surface has high nonlinearity, and cannot be represented as a Fourier or Taylor series with a reasonable number of coefficients for desired accuracy [15]. Thus, the best option is to use a lookup table for the current reference surface as well as for the torque and flux linkage surfaces.

Consider some small positive angle (point O in Figure 3a) where the positive torque can be produced by phases A and C. Set the initial phase A and C currents to zero. The phase C current can be adjusted by integral torque controller, which has some gain k as shown in Figure 5a. The integral torque controller regulates the phase C current until the error in the torque reaches the permitted tolerance. Thereafter, the phase A current starts to increase (see Figure 5b). The torque controller maintains its operation, continuously time varying the phase C current and keeping the total torque from both phases A and C close to the reference.

From the moment when the phase A current starts to grow, the system begins to track the minimum total current as it determines the ohmic losses in the motor windings. In the end of the transient from Figure 5b, the outputs hold the most efficient references of the phase currents $i_{a\,ref}$ and $i_{c\,ref}$, which can be used as points on the current reference surfaces for each phase as shown in Figure 3b.

A similar approach can be used for the angles when the positive torque is generated by one phase only. The only difference is that the search for minimal loss points is no longer needed due to the fact that total torque is produced by a single phase, and there is no extra degree of freedom.

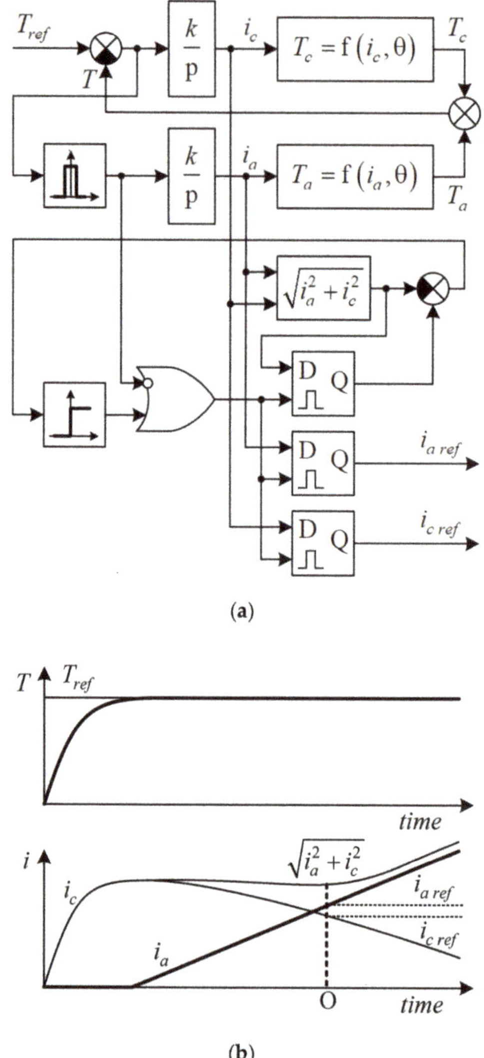

(a)

(b)

Figure 5. Evaluation of the current references for the given torque command: (**a**) block diagram, (**b**) transient during optimal reference evaluation.

4.3. Control System Flowchart

The control system is divided into an offline time-consuming algorithm placed into the initialization stage, as shown in Figure 6a, and a real-time control interrupt (see Figure 6b), which implements the proposed model predictive control strategy.

First, for the given representation of the magnetization surface, the initialization procedure evaluates the flux linkage surface for the desired resolution of the two-dimensional array, for example, 100·100 points. Next, these data are used to evaluate the torque surface using Equation (12). Finally, the most time-consuming algorithm is executed, which evaluates current references with respect to the range of all possible torque commands and angles using the iterative algorithm shown in Figure 5. Thereafter, the control system can start operation and rotate the machine.

The interrupt routine, which is shown in Figure 6b, should be executed at the end of the PWM cycle. The control system reads the feedback from the current, voltage, and position sensors inside this interrupt. At first, it uses the current reference surface to obtain current references for the next computed rotor angular position using Equation (5). Then, it obtains flux linkages for the current and next referenced steps from the magnetization surface. Finally, the voltage command and the duty cycles are evaluated and applied as the references to the PWM generator just before the new PWM cycle starts.

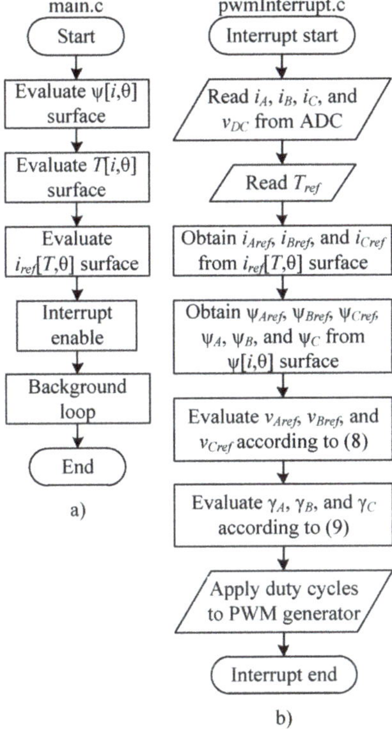

Figure 6. Flowchart of the control system: (**a**) initialization procedure and background loop, (**b**) interrupt with MPC strategy.

5. Simulation Results

There are various approaches to building a model of SRM. One of the most convenient options is to use a linearized magnetization profile [16], as shown in Figure 7, which allows fast simulation with simple equations for torque estimation to be undertaken. Phase inductance below the saturation knee was found from the following equation:

$$L = L_{av} - \Delta L \cos\theta, \tag{13}$$

where L_{av} is the average inductance, ΔL is the half-sum of maximum and minimum inductances, and θ is the rotor electrical angular position. When operating above saturation current i_{sat} the phase differential inductance becomes equal to the minimal inductance of the motor for a misaligned position as assumed in Figure 7.

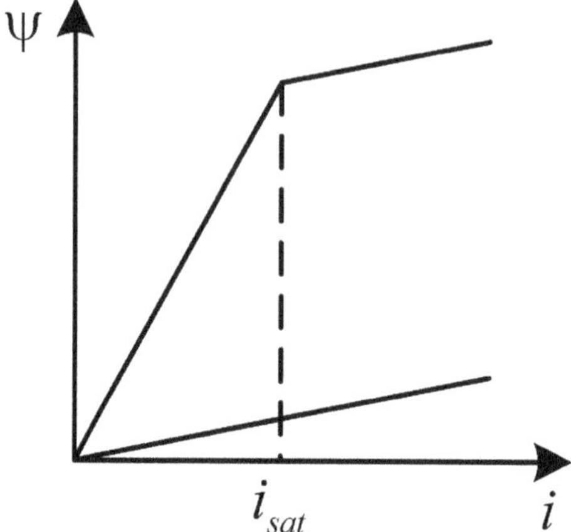

Figure 7. Linearized magnetization profile of SRM model.

The Euler integration method was used to obtain flux linkage in each phase and the resulting equation is given below:

$$\psi[m] = \psi[m-1] + (v - iR)h,\tag{14}$$

where $\psi[m]$ and $\psi[m-1]$ are the new and previous flux linkages of the motor model, respectively, and h is the integration step size.

The power converter, which contains an asymmetrical H-bridge, is depicted in Figure 1. It can only produce a positive current in any phase, which should be taken into account in the model when zero or negative voltage is applied to the winding. If the numerically integrated flux linkage value in the model becomes negative, then it should be set to zero:

$$\psi[m] = \begin{cases} \psi[m], \ \psi[m] \geq 0; \\ 0, \ \psi[m] < 0. \end{cases}\tag{15}$$

The value of the current in the phase winding was calculated in accordance with the estimated flux linkage and the inductance value found from Equation (13):

$$i = \frac{\psi}{L}.\tag{16}$$

If the evaluated current lies above the saturation knee, which can be checked by comparing it with the saturation current, then Equation (13) is not applicable and the phase current should be evaluated taking into account the differential inductance of the phase using the following equation:

$$i = i_{sat} + \frac{\psi - L \cdot i_{sat}}{L_{min}}\tag{17}$$

instead of Equation (16), where L_{min} is the minimum or differential inductance of the winding. If the actual current is smaller than the saturation current value, then the torque of a single phase was calculated using:

$$T = \frac{i^2}{2}\frac{dL(\theta)}{d\theta} = \frac{i^2}{2}\frac{d(L_{av} - \Delta L \cos\theta)}{d\theta} = \frac{\Delta L}{2}i^2 \sin\theta.\tag{18}$$

The torque in the operation point above the saturation knee, in addition, was evaluated using co-energy, which can be expressed as:

$$W'(i, \theta) = (L_{av} - \Delta L \cos \theta)\left(\frac{i_{sat}^2}{2} + i_{sat}(i - i_{sat})\right) + L_{min}i_{sat}(i - i_{sat}). \tag{19}$$

Then torque was found using:

$$T = \left.\frac{\partial W'(i, \theta)}{\partial \theta}\right|_i = \left(i_{sat}i - \frac{i_{sat}^2}{2}\right)\Delta L \sin \theta. \tag{20}$$

In this research, simulation was performed using C++ Builder. This tool was utilized in order to produce code for a control system suitable for further implementation using a microcontroller. The simulation was performed with the integration step size equal to 0.1 us. The parameters of the drive are listed in Table 1. As the control strategy contains only lookup table operation, it was supposed that the computation time is close to zero and that it is possible to evaluate the voltage commands for the next PWM cycle immediately at the end of the current PWM period.

Table 1. Switched reluctance drive (SRD) parameters.

Parameter	Value	Units
Minimal inductance	10.0	mH
Maximum inductance	100.0	mH
Saturation current	20	A
Phase resistance	0.05	Ohm
Pole pairs	4	-
Maximum phase current limit	100	A
DC link voltage	600	V
PWM frequency	2.0	kHz

The motor was running at 80 rad/s. The torque reference was changed during simulation as shown in Figure 8a. Initially it was set to 30 Nm, then after 30 PWM cycles it was set to 10 Nm, and finally to 45 Nm after 60 PWM cycles. The torque response transient time was strongly dependent on the current inductance of the winding. For example, at the beginning of the simulation, the torque slope was limited by the high unsaturated inductance of phase C. As the current reached the saturation knee, the torque on the shaft started to grow more quickly. Then, inside the saturated region, the phase currents and output torque were regulated more rapidly. At the end of commutation cycle of the phase, the current reference varied faster than the minimum possible current slope due to the voltage limit.

The torque stabilization is sufficiently precise over the entire range of the loads. It has visible deviations due to the PWM nature of the inverter, which can be minimized by increasing the PWM frequency. The same load profile was applied to the motor operating at a higher speed equal to 480 rad/s (see Figure 8b). With the growth of the electrical speed, the voltage limit made stabilization of the commanded current impossible. Therefore, torque pulsation appears. This behavior is natural for switched reluctance drives [17,18] and can be slightly improved by adjusting the current reference profile with respect to the voltage limit.

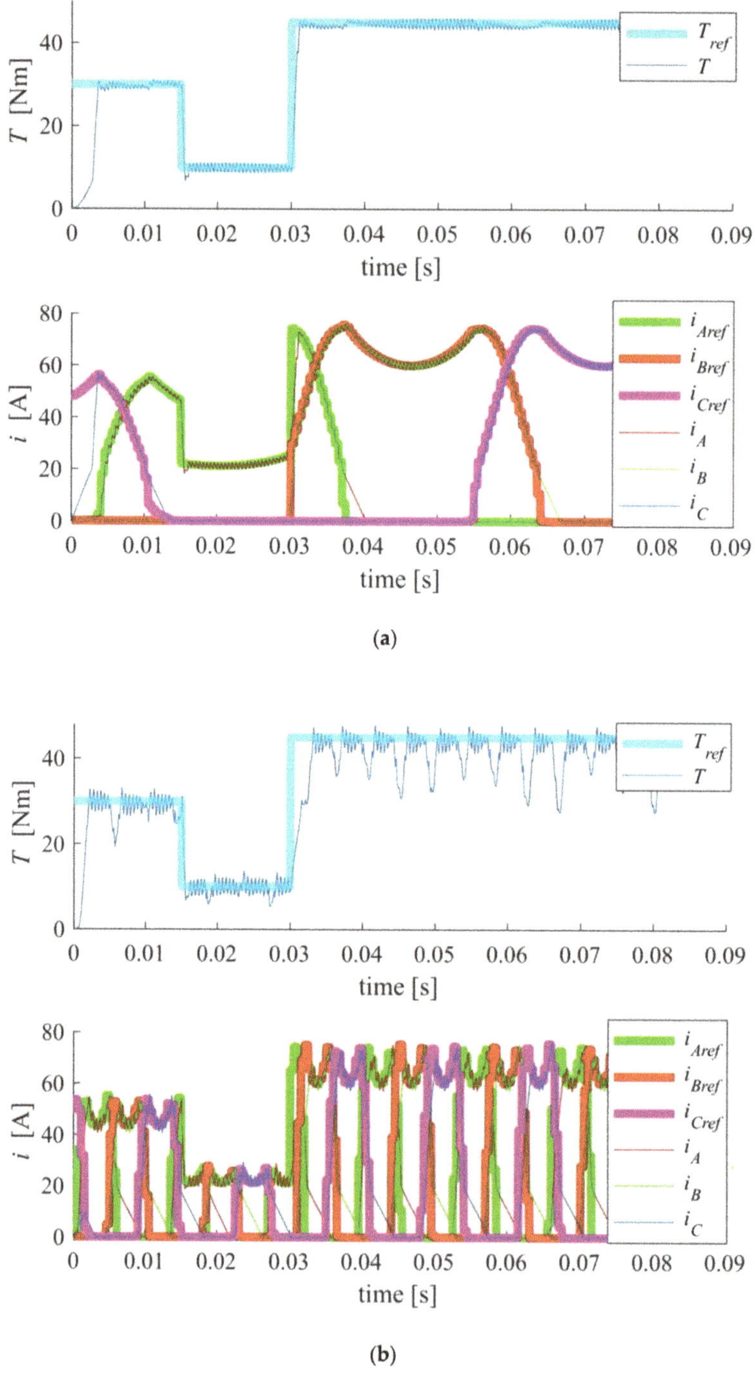

Figure 8. Operation at constant electrical speed: (**a**) 80 rad/s—mostly absent of voltage constraints; (**b**) 480 rad/s—operation with regular voltage constraints.

6. Conclusions

Although the problem of precise torque control in switched reluctance motor drives was partially solved by means of finite control set model predictive control, the drive still suffers from an unpredictable switching rate and acoustic noise. The proposed solution implements a continuous control set MPC controlling power converter using pulse-width modulation and requires small computation efforts. It operates utilizing the assumption that optimal current reference profiles for each torque reference and angular position can be evaluated offline from the magnetization surface of the electrical machine. By knowing the current reference and magnetization surface, the voltage commands for the PWM-driven inverter can be evaluated using simple lookup tables, which has the advantage of practically zero computation delay and allows the calculation of the control law immediately at the end of each PWM cycle before applying the new duty cycles for the next PWM period.

The proposed CCS MPC can be implemented using modern microcontrollers. It was verified by a simulation model where accurate torque stabilization was achieved from zero to the rated speed. The settling time was limited by the supply voltage and the phase inductance.

Future work will be devoted to the practical implementation and verification of the proposed control strategy. In addition, it will address several questions that arose during this research.

Due to the current and torque slope limit when operating below the saturation knee, the proposed current reference profile is not optimal in terms of settling time. Nonetheless, it provides good efficiency of the motor and drive. Similar optimization of the current reference profiles is possible for operation at high speeds considering the voltage limit.

The sensitivity of the method to the inaccuracies of the model should be investigated, as well as the ability to use the current tracking error for identification of the rotor angular position for encoderless control.

Author Contributions: General idea, A.A.; Simulation software, A.Z. and A.A.; Simulation model verification, C.H.; Methodology, V.Š.; Simulation and data analysis, G.L.D. and A.B.; Validation, A.B.; Writing original draft, A.A. and G.L.D.; Writing review & editing, G.L.D., C.H. and V.Š. All authors have read and agreed to the published version of the manuscript.

Funding: This work was financially supported by Government of Russian Federation, Grant 08-08.

Conflicts of Interest: The authors declare no conflict of interest.

Nomenclature

AC	Alternating Current
ADC	Analog-to-Digital Converter
CCS	Continuous Control Set
CMPR	Compare Period Register
DTC	Direct Torque Control
FCS	Finite Control Set
FOC	Field-oriented Control
FPGA	Field-Programmable Gate Array
IGBT	Insulated-Gate Bipolar Transistor
MPC	Model Predictive Control
PWM	Pulse-Width Modulation
QEP	Quadrature Encoded Pulses
SRD	Switched Reluctance Drive
SRM	Switched Reluctance Machine
TPR	Timer Period Register
VS	Voltage Sensor

References

1. Rassolkin, A.; Kallaste, A.; Vaimann, T.; Tiismus, H. Control Challenges of 3D Printed Switched Reluctance Motor. In Proceedings of the 2019 26th International Workshop on Electric Drives: Improvement in Efficiency of Electric Drives (IWED), Moscow, Russia, 30 January–2 February 2019.
2. Andrada, P.; Blanqué, B.; Martínez, E.; Perat, J.I.; Sánchez, J.A.; Torrent, M. Design of a Novel Modular Axial-Flux Double Rotor Switched Reluctance Drive. *Energies* **2020**, *13*, 1161. [CrossRef]
3. Mademlis, C.; Kioskeridis, I. Performance Optimization in Switched Reluctance Motor Drives With Online Commutation Angle Control. *IEEE Trans. Energy Convers.* **2003**, *18*, 448–457. [CrossRef]
4. Yu, C.; Xiao, Z.; Huang, Y.; Zhu, Y. Two-step commutation control of switched reluctance motor based on PWM. *J. Eng.* **2019**, *2019*, 8414–8418. [CrossRef]
5. Zhang, M.; Bahri, I.; Mininger, X.; Vlad, C.; Xie, H.; Berthelot, E. A New Control Method for Vibration and Noise Suppression in Switched Reluctance Machines. *Energies* **2019**, *12*, 1554. [CrossRef]
6. Huang, L.; Zhu, Z.Q.; Feng, J.H.; Guo, S.Y.; Li, Y.F.; Shi, J. Novel Current Profile of Switched Reluctance Machines for Torque Density Enhancement in Low-Speed Applications. *IEEE Trans. Ind. Electron.* **2019**, *46*, 1–11. [CrossRef]
7. Ye, J.; Bilgin, B.; Emadi, A. An Offline Torque Sharing Function for Torque Ripple Reduction in Switched Reluctance Motor Drives. *IEEE Trans. Energy Convers.* **2015**, *30*, 726–735. [CrossRef]
8. Lukman, G.F.; Nguyen, X.S.; Ahn, J.-W. Design of a Low Torque Ripple Three-Phase SRM for Automotive Shift-by-Wire Actuator. *Energies* **2020**, *13*, 2329. [CrossRef]
9. Inderka, R.B.; Doncker, R.W.A.A. De DITC—Direct Instantaneous Torque Control of Switched Reluctance Drives. *IEEE Trans. Ind. Appl.* **2003**, *39*, 1046–1051. [CrossRef]
10. Krishnan, R. *Switched Reluctance Motor Drives Modeling, Simulation, Analysis, Design, and Applications*, 1st ed.; CRC Press: Boca Raton, FL, USA, 2001; ISBN 0849308380.
11. Parreira, B.; Rafael, S.; Pires, A.J.; Branco, P.J.C. Obtaining the magnetic characteristics of an 8/6-switched reluctance machine: FEM analysis and experimental tests. *IEEE Trans. Ind. Electron.* **2005**, *52*, 1635–1643. [CrossRef]
12. Ramanarayanan, V.; Venkatesha, L.; Panda, D. Flux-linkage characteristics of switched reluctance motor. In Proceedings of the International Conference on Power Electronics, Drives and Energy Systems for Industrial Growth, New Delhi, India, 8–11 January 1996; Volume 1, pp. 281–285.
13. Lin, Z.; Reay, D.S.; Zhou, B. Experimental measurement of Switched Reluctance Motor non-linear characteristics. In Proceedings of the IECON 2013—39th Annual Conference of the IEEE Industrial Electronics Society, Vienna, Austria, 10–13 November 2013; pp. 2827–2832.
14. Anuchin, A.; Grishchuk, D.; Zharkov, A.; Prudnikova, Y.; Gosteva, L. Real-time model of switched reluctance drive for educational purposes. In Proceedings of the 2016 57th International Scientific Conference on Power and Electrical Engineering of Riga Technical University (RTUCON), Riga, Latvia, 13–14 October 2016.
15. Cossar, C.; Popescu, M.; Miller, T.; McGilp, M. On-line phase measurements in switched reluctance motor drives. In Proceedings of the 2007 European Conference on Power Electronics and Applications, Aalborg, Denmark, 2–5 September 2007; pp. 1–8.
16. Li, C.; Wang, G.; Liu, J.; Li, Y.; Fan, Y. A Novel Method for Modeling the Electromagnetic Characteristics of Switched Reluctance Motors. *Appl. Sci.* **2018**, *8*, 537.
17. Ding, W.; Liu, G.; Li, P. A hybrid control strategy of hybrid-excitation switched reluctance motor for torque ripple reduction and constant power extension. *IEEE Trans. Ind. Electron.* **2020**, *67*, 38–48. [CrossRef]
18. Rekik, M.; Besbes, M.; Marchand, C.; Multon, B.; Loudot, S.; Lhotellier, D. Improvement in the field-weakening performance of switched reluctance machine with continuous mode. *IET Electr. Power Appl.* **2007**, *1*, 785–792. [CrossRef]

Article

Additive Manufacturing and Performance of E-Type Transformer Core

Hans Tiismus [1,*], Ants Kallaste [1], Anouar Belahcen [2], Anton Rassolkin [1], Toomas Vaimann [1] and Payam Shams Ghahfarokhi [1,3]

[1] Institute of Electrical Power Engineering and Mechatronics, Tallinn University of Technology, 19086 Tallinn, Estonia; ants.kallaste@taltech.ee (A.K.); anton.rassolkin@taltech.ee (A.R.); toomas.vaimann@taltech.ee (T.V.); payam.shams@taltech.ee (P.S.G.)

[2] Department of Electrical Engineering and Automation, Aalto University, 02150 Espoo, Finland; Anouar.Belahcen@aalto.fi

[3] Department of Electrical Machines and Apparatus, Riga Technical University, Kaļķu iela 1, LV-1658 Riga, Latvia

* Correspondence: hans.tiismus@taltech.ee

Abstract: Additive manufacturing of ferromagnetic materials for electrical machine applications is maturing. In this work, a full E-type transformer core is printed, characterized, and compared in terms of performance with a conventional Goss textured core. For facilitating a modular winding and eddy current loss reduction, the 3D printed core is assembled from four novel interlocking components, which structurally imitate the E-type core laminations. Both cores are compared at approximately their respective optimal working conditions, at identical magnetizing currents. Due to the superior magnetic properties of the Goss sheet conventional transformer core, 10% reduced efficiency (from 80.5% to 70.1%) and 34% lower power density (from 59 VA/kg to 39 VA/kg) of the printed transformer are identified at operating temperature. The first prototype transformer core demonstrates the state of the art and initial optimization step for further development of additively manufactured soft ferromagnetic components. Further optimization of both the 3D printed material and core design are proposed for obtaining higher electrical performance for AC applications.

Keywords: additive manufacturing; soft magnetic materials; selective laser melting; iron losses; magnetic properties; transformer

Citation: Tiismus, H.; Kallaste, A.; Belahcen, A.; Rassolkin, A.; Vaimann, T.; Shams Ghahfarokhi, P. Additive Manufacturing and Performance of E-Type Transformer Core. *Energies* **2021**, *14*, 3278. https://doi.org/10.3390/en14113278

Academic Editor: Salvatore Musumeci

Received: 5 May 2021
Accepted: 31 May 2021
Published: 3 June 2021

Publisher's Note: MDPI stays neutral with regard to jurisdictional claims in published maps and institutional affiliations.

1. Introduction

Metal additive manufacturing (AM) is maturing, enabling previously unavailable production possibilities in terms of feasible product complexity and personalization. As currently, the cost per part of AM is still relatively high, it has been most applicable for parts for high tech industries: producing specialized parts benefiting the most from the topology optimization possibilities of AM. For example, 3D printing has been utilized for the production of more efficient and long-lasting inductor coils [1], stronger, cheaper and lighter aircraft fuel nozzles [2], and high performance heat exchangers [3].

In parallel to the printing of structural, thermal, and electrical components, research interest in printed soft magnetic materials and topology optimized electromechanical components has spiked drastically over recent years. It has been proposed that with the easily available computational power and free-form printing capabilities of AM systems, next generation electrical machine designs could be modelled and constructed by the research community. These topology optimized designs (with reduced weight, integrated cooling channels, reduced inertia, increased heat exchange etc.) could be prototyped in-house, significantly reducing the lead time, cost, and machinery involved [4].

State of the art additive manufacturing of electromagnetic devices involves selective laser melting (SLM) printing of conductive and soft magnetic materials with air gaps partitioning the material structure for separating individual turns in coils and reducing the induced eddy currents in soft magnetic cores [5,6]. The air gaps are printed due to the current lack of multi-material printing capacity of SLM systems, limiting the parallel printing of conductive, core, and insulation materials. The introduction of airgaps considerably reduces the power density of the components, however, as gapped printed component fill factor is typically relatively low (in the range of 60%) [6,7].

Despite extensive material optimization of different soft magnetic alloys, relatively few functional components or devices have actually been printed and characterized. For this reason, in this work, a full small-scale transformer core is printed, characterized, and compared with a commercial transformer. The simplistic design of an E-type transformer makes it ideal for the next step of testing additively manufactured magnetic material capacity and performance for electrical machine applications (succeeding the characterization of small-scale toroidal samples). In this paper, a novel interlocking core design is employed for eddy current reduction, which exhibits a competitive component fill factor. The paper is divided into two larger sections. The first part describes the 3D printed core design and its fabrication process, and the second the characterization and comparison of the printed core with conventional cores.

2. Transformer Core Design

2.1. Commercial Transformer

The 3D printed core design investigated in this paper was based on the commercially available 30 VA single phase isolation transformer provided by MS Balti Transformers Ltd (Tallinn, Estonia). The transformer was chosen based on its suitable size, type, and availability. Its shell-type transformer core is constructed from E-type stampings of grain-oriented M 165-35S silicon steel. The conventional transformer design with its dimensions are detailed on Figure 1.

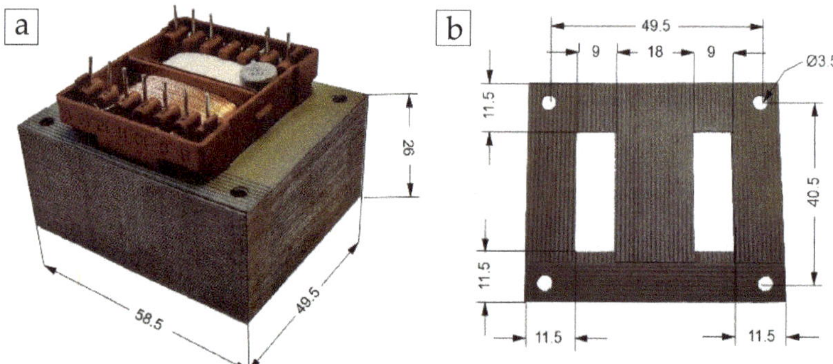

Figure 1. Investigated conventional transformer: (**a**) Core dimensions, (**b**) E-type stamping dimensions in detail.

The fully encapsulated modular windings of the transformer are utilized in both the conventional and 3D printed core designs. The modular windings are incorporated in both designs in order to improve the comparability of the transformer core performance and to demonstrate the compatibility of 3D printed and conventional parts. The nominal parameters of the windings are characterized in Table 1.

2.2. 3D Printed Design

Next, an SLM printing system was utilized for the 3D printing of the full transformer core. The 3D printed core design was required to exhibit compatibility with the modular windings, incorporate the segregated structure for classical eddy current loss reduction with high filling factor, and adhere to the printing system requirements. Lamination thickness of

0.95 mm was chosen to obtain high fill factor and mechanical strength of the first prototype. For all segregated designs considered, it was critical to achieve continuous geometries (with minimal air gaps dividing the flux paths) with maximal flux path cross sectional area (high fill factor). Furthermore, since the printed transformer must be comprised of at least two parts (to accommodate the modular winding), optimization of the inter-part air gap must be considered. In conventional transformers, the influence of the inter-stamping airgaps is typically reduced by overlapping stamping layers: which facilitates the flux paths through the adjoining stampings. Similar overlap between the flux-guides can be realized in printed designs.

Table 1. Nominal parameters of the modular transformer coil.

Winding	Turns	Resistance (Ω)	Nominal Voltage (V)	Nominal Current (A)	Insulation Class
Primary	1370	98	230	0.17	H
Secondary 1	151	1.35	25.1	1.3	F
Secondary 2	56	2.7	9.3	0.25	F

For simplicity, in this paper, only conventional stamping inspired designs were considered for 3D printing. In Figure 2, three considered transformer core designs are illustrated: (a) a laterally laminated interlocking design from four parts, (b) an axially laminated gapped design from two parts, and (c) an axially laminated interlocking design from four parts. The axially laminated interlocking design was chosen for printing due to its simplicity and similarity to the conventional design, its high achievable fill factor and its post-processing possibilities: all of the unmelted powder can be removed between the laminations post-printing and, if needed, all of the surfaces can be cleaned and oxidized or varnished for enhanced inter-lamination electrical resistance.

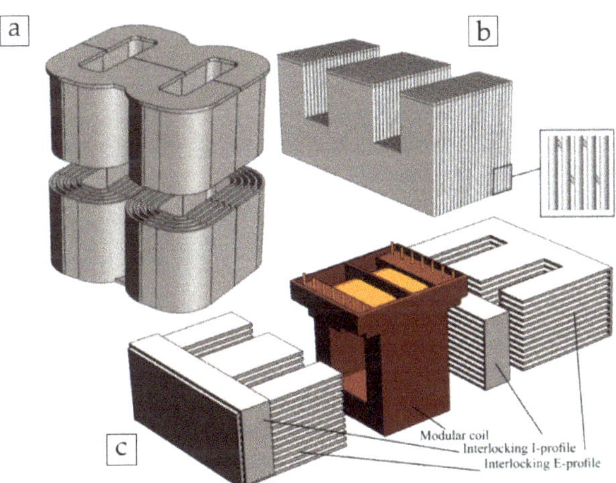

Figure 2. Considered lamination strategies: (**a**) Laterally laminated interlocking design, (**b**) Axially laminated design with air-gapped core structure, (**c**) Exploded view of the 3D printed transformer core design with interlocking axial laminations comprising four individual components.

3. Methods

3.1. Powder Characteristics

Transformer parts were printed with identical powder, processing, and annealing parameters to the previous study characterizing the AC and DC losses of the printed material [6]. Pre-alloyed, gas-atomized Fe-Si provided by Sandvik group was utilized for printing. The powder exhibited roughly spherical particle shape with a median diameter of 38 μm, and its chemical composition is described in Table 2. The powder size, shape, and chemical composition were verified to verify the manufacturer declaration.

Table 2. Chemical composition of the employed Fe-Si powder.

Elements	Fe	Si	Mn	Cr	Ni	C
Wt%	Balance	3.7	0.2	0.16	0.020	0.01

3.2. SLM Printing of the Transformer Core

Transformer core parts were printed on the SLM Solutions GmbH Realizer SLM-280. The printing system provides a 280 × 280 × 350 maximum build envelope and a single 1070 nm yttrium scanning laser (1 × 700 W). Custom smaller build platform (D100 mm) and re-coater were used for printing of the transformer core, designed for streamlining the powder substitution between projects for different raw powders.

Laser re-melting strategy was used to prevent the powder balling related uneven growth of the relatively large transformer parts during printing, which can result in rough porous material structure or the termination of the print job due to re-coater jamming. The phenomenon is related to an oxide film on the preceding layer impeding interlayer bonding and leading to balling, due to insufficient wetting of the molten metal on the oxide layer [8]. The balling phenomenon can be reduced in a higher purity environment (oxygen level below 0.1%), applying a combination of high laser powder and low scanning rate or applying re-melting scanning on the part [9].

Stripe (10mm wide) scan pattern was utilized with 30° rotation between layers. All of the printing was conducted in a nitrogen inert gas environment because of its relatively low cost. Platform pre-heating was not utilized as the custom reduced platform is not equipped for it. A summary of the main laser printing parameters is presented in Table 3.

Table 3. Summary of the printing parameters.

Parameter	Value
Layer thickness	50 μm
Hatch distance	120 μm
Laser Power	250 W (primary)/100 W (secondary)
Scanning velocity	0.5 m/s (primary)/0.5 m/s (secondary)
Scan strategy	Stripes
Environment	Nitrogen
Oxygen content	~0.1%

Transformer printing was completed in three parts in a total of 16 h: interlocking E-profiles separately (2 × 6 h) and the I-profiles in the same build (1 × 4 h). The printed components are illustrated in Figure 3: showing the surface finish, support structure, and the powder bed post-printing. Some concave warpage of the E-profiles was observed after separation from the build platform due to internal part stress, which obstructed the transformer assembly. Its causality can be traced to the relatively high internal stresses induced in part by the micro-welding process of SLM, and it can be resolved through the annealing of the printed parts at moderate temperature, pre-cutting from the platform for stress relief. Next, the support surfaces were polished and the inter-lamination air-gaps were lightly sanded for improved surface finish and fitting of the components.

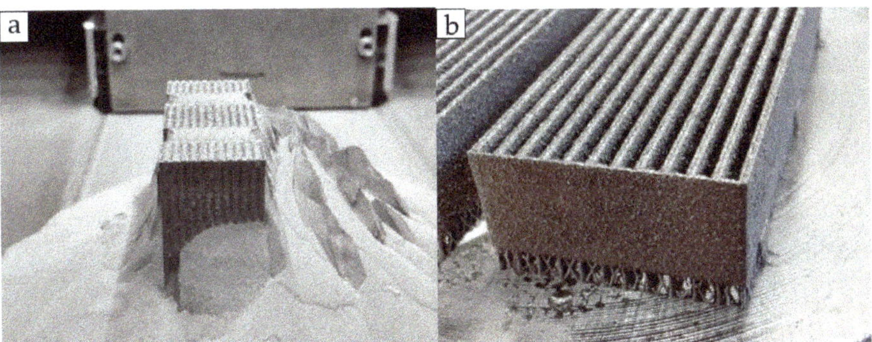

Figure 3. SLM printed transformer components: (**a**) E-profile component post-printing, (**b**) I-profile component welded on the baseplate.

3.3. Annealing

After mechanical post-processing, the printed transformer parts were annealed at 1200 °C in a low vacuum environment (~0.1 mBar) with a heating rate of 300 K/h, maintained at the target temperature for 1 h and then slowly furnace-cooled to room temperature.

3.4. Material Properties

The additively manufactured 3.7% silicon steel shows comparable magnetic performance to non-oriented conventional silicon steels after thermal treatment. Magnetization of 1.5 T is achieved at 1800 A/m, exhibiting electrical resistivity of 56.9 $\mu\Omega\cdot$cm and hysteresis losses of 0.61 ($W_{10,50}$) and 1.7 ($W_{15,50}$) W/kg [6]. In comparison, a typical non-oriented steel M235-35A used for electrical machine fabrication exhibits total core losses of 0.92 ($W_{10,50}$) and 2.35 W/kg ($W_{16,50}$), resistivity of 59 $\mu\Omega\cdot$cm, and magnetization of 1.53 T at 2500 A/m. In this paper, we are comparing the additively manufactured core with a conventional Goss textured silicon steel M165-35S (equivalent to M111-35N) core, which shows superior magnetic properties to the non-oriented materials for transformer applications, as presented on Figure 4. The grain-oriented transformer steel shows approximately 0.3 T greater saturation magnetization than both of the non-oriented steels.

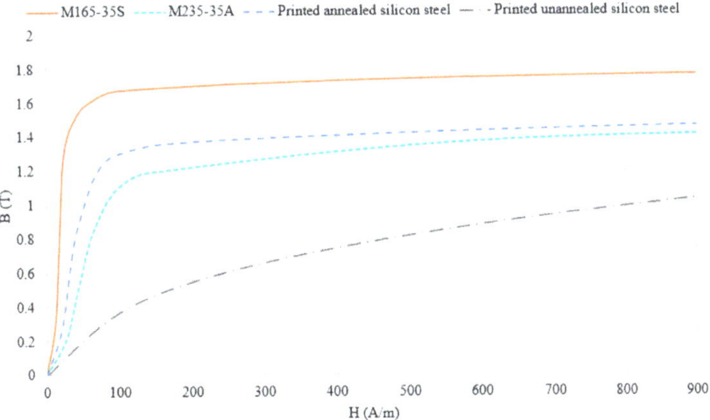

Figure 4. Magnetization curves of the studied materials: grain-oriented silicon steel M165-35S [10,11], non-oriented silicon steel M235-35A [12], printed annealed 3.7% silicon steel, and printed unannealed 3.7% silicon steel [6].

The grade designation of M165-35S of the Goss textured steel specifies 1.65 W/kg losses at 1.7 T ($W_{17,50}$), and a lamination thickness of 0.35 mm. The materials' exact silicon content, resistivity, and other typical properties are unspecified and depend on the manufacturer (manufacturing freedom in the range of grade specifications).

3.5. Transformer Characterization

The nominal performances of both the 3D printed and the conventional magnetic core transformers were characterized through open circuit and full load testing. The nominal parameters of the conventional transformer were obtained from the manufacturer's declaration. A drop in the nominal voltage is expected for the printed transformer due to its reduced fill factor, possible fitting defects (air-gaps between laminations), and lower saturation magnetization of the printed material. Its nominal voltage and iron losses were determined from the open circuit tests of the conventional transformer. To determine the transformer efficiencies, a load test was performed, where the transformer was energized up to nominal power. For thermal performance assessment, steady-state thermal images of the fully loaded transformers were captured with a Fluke Ti10 Thermal Camera.

The open circuit test setup is described in Figure 5, consisting of an autotransformer for variable voltage input and digital multimeters for measuring the voltage, current, and active power consumed in the transformer coil. In the open circuit test, the current drawn by the transformer establishes the magnetic field in the core. The active power consumed by the transformer signifies its total power loss, consisting mainly of magnetizing, and some ohmic, losses. The magnetizing losses summarize the energy lost from each magnetizing cycle, which are classically segregated into the hysteresis, classical, and excess eddy current loss.

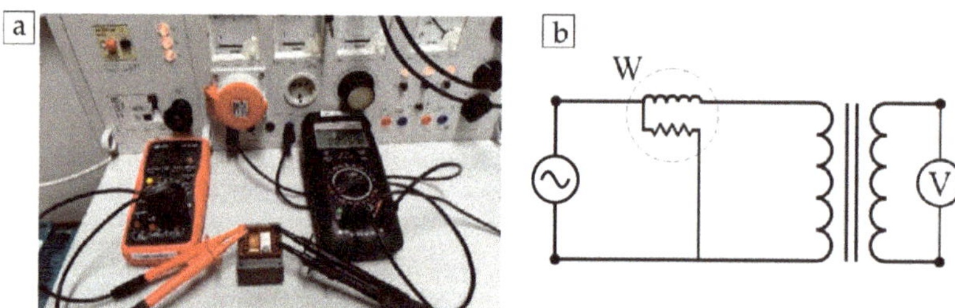

Figure 5. Open circuit transformer: test setup (**a**) and its schematic (**b**).

The ohmic losses are induced from joule heating of the coils due to the magnetizing current drawn. The total specific transformer core losses can be calculated from (1), where W is the active power loss measured in the open circuit test, I is the magnetizing current, R is the magnetizing coil resistance, and m is the weight of the core.

$$P = \frac{(W - IR^2)}{m} \tag{1}$$

Traditionally, the magnetic material loss behaviour is discussed in terms of cycle peak polarization (B_{max}) of the core. Unlike in the toroidal cores for magnetic material characterization [6,13], however, the flux density in the investigated transformer core can only be evaluated as an approximation, due to its uneven flux distribution. The analytical expression for calculating the peak polarization in a transformer can be derived from the differential form of Faraday's law (2), where E is the induced electromotive force by the switching magnetic field, N is the number of turns on the primary coil (1370), f is the excitation frequency of the magnetic field (50 Hz), B_{max} is the peak material polarization, S is the core cross sectional area, F is the core filling factor, U is the applied voltage on the primary coil, and U_r is the voltage drop over the primary coil.

$$E = N\frac{d\Phi}{dt} \rightarrow E_{max} = N2\pi fSFB_{max} \rightarrow B_{max} = \frac{E_{max}}{N2\pi fSF} = \frac{U - U_r}{N2\pi fSF} \quad (2)$$

Alternatively, the approximate material polarization can be evaluated from the material B-H curve (as presented on Figure 4) or by the finite element method (FEM) simulation. In both methods, the actual B-H curve of the transformer core can differ from the previously characterized material, most prominently due to air-gap related curve shearing. For B_{max} evaluation, the magnetic field strength in the transformer is calculated from (3), where N is the number of turns on the primary coil, i is the peak magnetizing current and l is the length of the mean magnetic flux path of the core. All FEM simulations are performed in open source finite element analysis software package Finite Element Method Magnetics (FEMM). The model accounts for the transformer cross sectional geometry, magnetized up to the peak magnetizing current measured from the open circuit test, including the material magnetization curve and fill factor, but excluding any gaps in the core internal structure.

$$H = \frac{Ni}{l} \quad (3)$$

4. Results

4.1. Assembled Transformer

The conventional and finished assembled printed transformer cores are presented in Figure 6. The overall transformer core dimensions correlated well, with the printed transformer exhibiting a slightly thinner and lighter core. The fill factor of the 3D printed core was measured from the axial centerline of the interlocking E-cores. For the conventional transformer, the fill factor was adopted from the stamping datasheets. The physical comparison of the transformer cores is presented in Table 4. No additional oxidation, treatment, or varnishing was applied to the surfaces of the 3D printed transformer core for increased eddy current reduction—the insulation is provided by the high natural surface roughness of the printed parts.

Figure 6. Printed (**a**) and conventional (**b**) transformer cores.

Table 4. Physical comparison of the transformer cores.

Core	Lamination Thickness (mm)	Fill Factor	Dimensions (mm)	Weight Core (kg)	Weight Coil (kg)	Varnish
Conventional	0.35	0.96	58.5 × 49.8 × 26.0	0.44	0.095	Yes
Printed	0.95	0.89	58.8 × 49.9 × 25.0	0.41	0.095	No

4.2. Performance

Open circuit tests of the transformers confirmed the flux drop in the core and the reduction of the sustainable operating voltage of the printed transformer. In Figure 7, both the magnetizing current drawn from the supply for generating the desired voltage and the iron loss behavior calculated from (1) are presented. At 40 mA magnetizing current, the conventional transformer is energized up to 230 V, while the printed transformer is energized to a 30% lower voltage of 160 V. This is due to the lower flux density sustained by the printed material. For energizing the printed transformer up to 230 V, a magnetizing current of 220 mA is required. This is inefficient, however, due to deep core oversaturation, requiring 450% more current than for magnetizing the conventional core and 30% more current than the rated full load current of the winding.

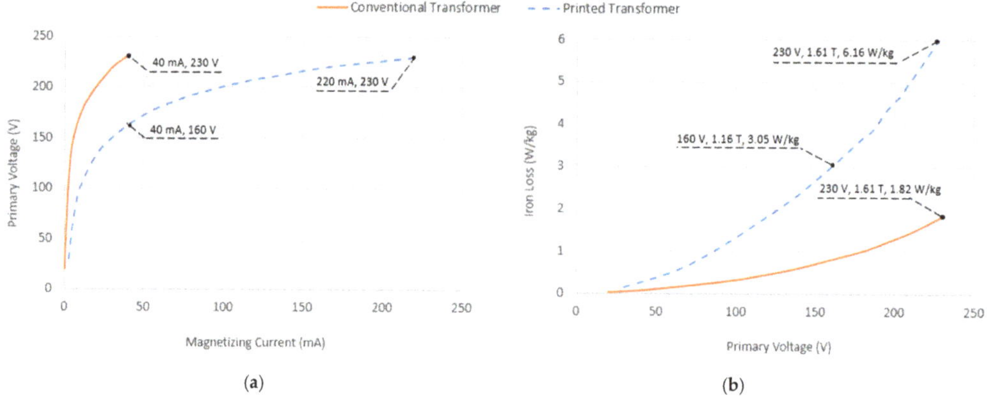

Figure 7. Magnetizing current drawn (**a**) and the specific core loss (**b**) of the tested transformers.

At 40 mA RMS excitation current (60 mA peak current), the analytically calculated (3) average H field generated in the magnetic core is 668 A/m, which corresponds to the magnetization of 1.72 T for M165-35S and 1.42 T for the annealed 3D printed material as determined from the magnetization curves in Figure 8. At 40 mA RMS excitation current, analogous excitation of both cores is achieved. Both are magnetized slightly above the approximate material knee-point and exhibit identical copper losses. Excitation of the conventional core to 160 V or the 3D printed core to 230 V would be impractical comparison-wise, as both states exhibit significantly differing magnetic behavior. At 160 V, the conventional core is still at the linear magnetic behavior: drawing only 7.6 mA magnetizing current and exhibiting 0.005 W of copper losses and 0.35 W of iron losses. At 230 V, the printed transformer shows deep saturation behavior, drawing 220 mA of magnetizing current, resulting in a significant voltage drop of 21.6 V, copper losses of 4.7 W, and iron losses of 2.6 W.

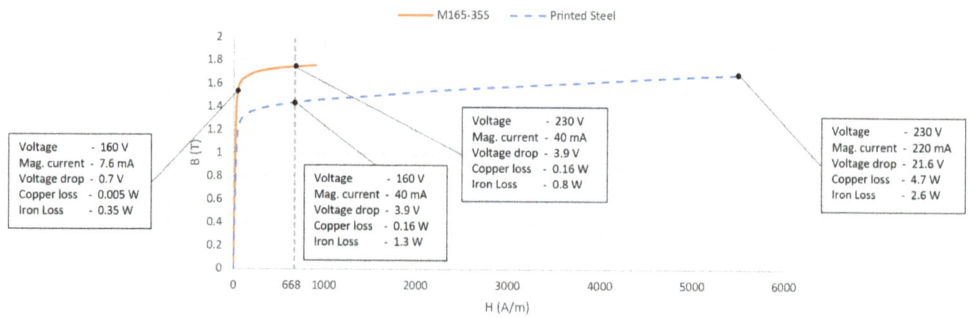

Figure 8. Core material magnetization curves correlated with the no load measurements of the investigated transformers.

FEM simulation of the transformer cores shows similar values of material magnetization: reaching 1.68 T for the conventional and 1.39 for the 3D printed core (Figure 9). Additionally, the simulation illustrates the uneven flux distribution in the core due to variations in transformer limb width. Analytical calculations with (2) show lower core flux density required for inducing a specific voltage in the core. For energizing the transformer up to 230 V, a flux density of 1.65 T is required, while for 160 V, a flux density of 1.26 T is required. The higher magnetization calculated from the experimental excitation current and FEM simulation is most likely the result of intra-lamination air-gaps, which shears the material magnetization curve and requires more current for achieving the same material polarization.

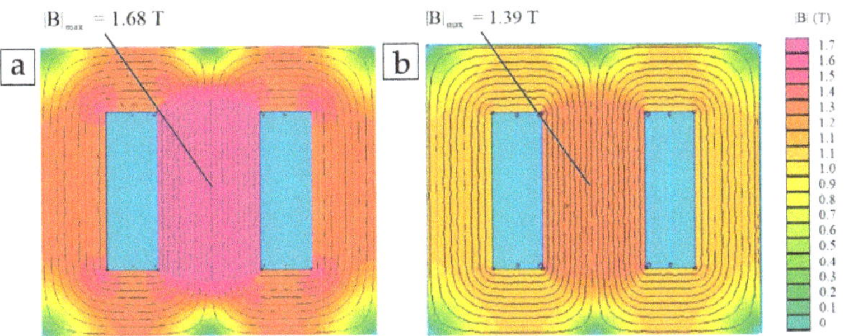

Figure 9. Flux distribution in the (**a**) conventional and (**b**) 3D printed transformer core.

Iron losses were identified as 1.82 W/kg for the conventional core at 230 V (at approximately 1.7 T, 50 Hz) and 3.05 W/kg for the 3D printed core at 160 V (in the range of 1.26–1.4 T, 50 Hz). Efficiency of the transformers was calculated from the load test measurements at both the ambient core temperature and the steady state temperature at full load conditions. The transformers reached steady state temperature after four hours of loading. The thermal images of the transformers are shown in Figure 10, with slightly higher heating observed for the 3D printed transformer core. The measured coil hotspot temperature was measured at 91.1 °C for the conventional core and at 95.1 °C for the 3D printed core. The core hotspots were measured with a thermocouple sensor due to the high reflectivity of the printed core, exhibiting temperatures of 71 °C (conventional) and 75 °C (3D printed).

At full load, the measured efficiency of the transformers ranged from 83.8% (21 °C) to 80.4% (71 °C) for the conventional transformer and 74.7% (21 °C) to 70.1% (75 °C) for the 3D printed transformer. The efficiency-load characteristic is presented in Figure 11. The highest efficiencies were measured at 41% load at ambient core temperature, reaching an efficiency of 88.7% for the conventional transformer and 80.5% for the 3D printed core. The efficiency of the 3D printed core was approximately 10% lower over the full measurement range. Due to the material saturation and inter-lamination air-gap related reduction of nominal voltage, the printed transformer core sustained reduced power density when compared to the conventional core. The transformer power density dropped 34% from 59 W/kg to 39 W/kg. The results of the transformer performance characterization are summarized in Table 5.

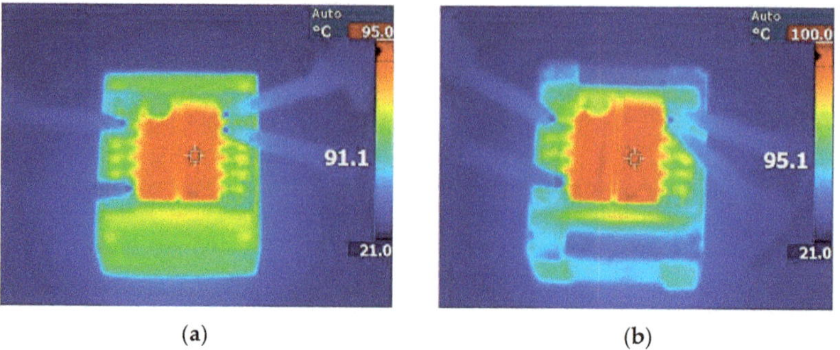

(a)   (b)

Figure 10. Steady state temperature of the studied transformers in the (**a**) conventional core and (**b**) 3D printed transformer core.

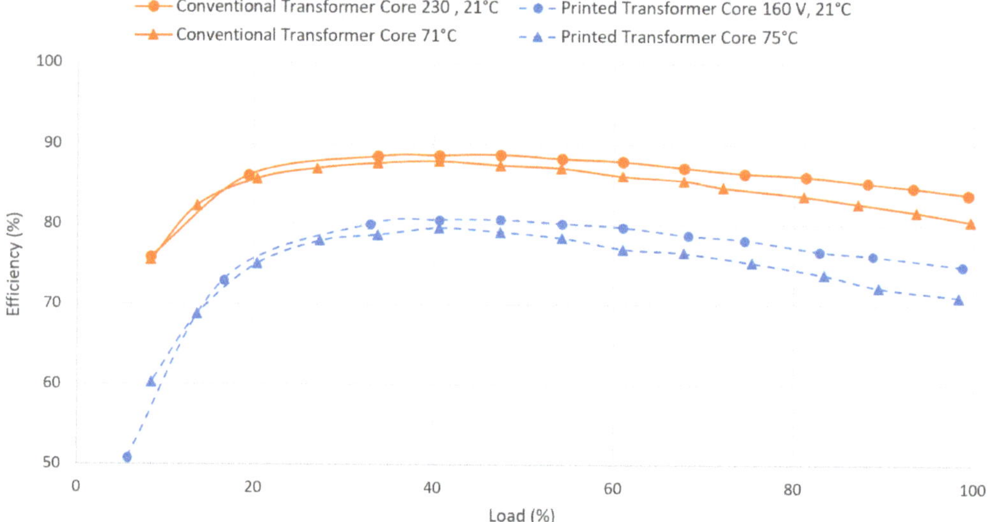

Figure 11. Efficiency-load characteristics of the studied transformers.

Table 5. Comparison of transformer performance.

Parameter (Full Load, 21 °C)	3D Printed Core	Conventional Core
Nominal Voltage	160 V	230 V
Nominal Current	0.17 A	0.17 A
Power Factor	0.97	0.97
Input Power	27.2 VA	39.1 VA
Output power	19.8 VA	31.8 VA
Efficiency (ambient temperature)	74.7%	83.8%
Efficiency (operating temperature)	70.1%	80.5%
Power Density (Core)	47 VA/kg	72 VA/kg
Power Density (Full Transformer)	39 VA/kg	59 VA/kg

5. Discussion

The characterized transformers show typical performance values for small 20–30 VA power rating single-phase transformers. From manufacturer datasheets, the typical efficiency for a 30 VA rated power transformer is in the range of 83 [14]–81% [15], which decreases to 77% [14] at 22 VA and to 65% [14] at 4.5 VA. The rated power densities vary significantly depending on the design (some designs are fully encased), and are typically in the range of 56 [15]–39 VA/kg [14] for 30 VA rated transformers and slightly lower (50 [16]–39 [14] VA/kg) for 20 VA rated transformers. In this study, we obtained an efficiency of 80.5% for the conventional transformer and 70.1% for the 3D printed transformer core at steady state temperature. The 10% reduced overall transformer efficiency can most prominently be attributed to the eddy currents generated in the 170% thicker laminations of the printed design. The reduced power density of the printed design can be attributed to both a larger degree of assembly defect related air-gaps within the core and the overall lower magnetic saturation of the printed material compared to the Goss textured conventional steel. Both designs are within the range of typical power density values for low power transformers.

The 3D printed core exhibited iron losses of 3.05 W/kg at 160 V transformer energization. Analytical calculations identify an average B_{max} of 1.26 T at this transformer voltage level. Comparing the magnetizing values with previously measured 3D printed material magnetization curves, its shearing is proposed. Due to the air-gaps in the assembled printed design, more magnetizing current is required for the same material polarization and voltage generated by the transformer. Similar iron loss values have been measured by Plotkowski et al. for a 3D printed E-type transformer core [17]. In their work, they achieved a core loss of 3.5 W/kg ($W_{10,60}$) at 1.0 T, 60 Hz magnetization for a printed 3% silicon steel lamination inspired core. They achieved considerably improved losses with more complex geometry, reaching approximately 1.5 W/kg ($W_{10,60}$) to 3.2 W/kg ($W_{15,60}$) with 'Hilbert pattern' 6% silicon steel. It is important to note, however, that in their work approximately 56% core fill factor was achieved, resulting in low power density and voltage generation of the transformer.

Further optimization of both the component topology and its material properties are unavoidable for achieving high performance 3D printed transformer cores. To obtain high magnetic polarization (high power density) of the printed material with minimal magnetomotive force, a higher degree of control of the printed material grain structure must be achieved. The effect of the grain structure orientation in relation to the magnetic field is significant as illustrated by Figure 12 [18]. In conventional stampings, the grain-oriented pronounced Goss texture can be achieved with various hot and cold rolling stages of the steel sheets. In printed material, the optimization of the material grain structure is largely immature, with some grain structure evolution observed in [13], in heat treated laser-remelted printed silicon steel samples.

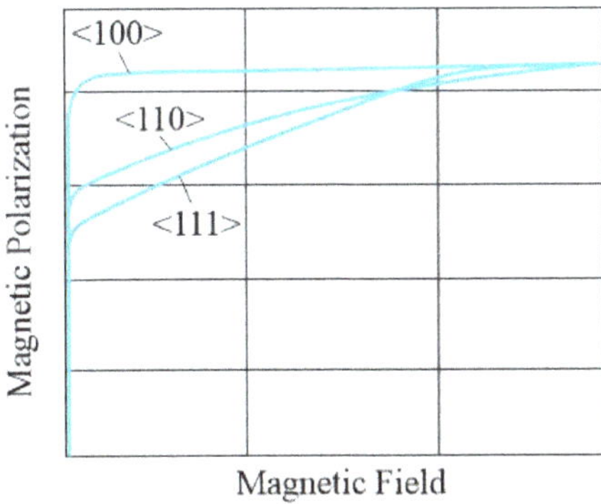

Figure 12. Polarization of the magnetic grains oriented in easy <100>, medium <110>, and hard <111> magnetization axis direction in relation to the magnetic field (in arbitrary units).

Several topological improvements can be applied to the transformer for enhanced performance. The printed transformer topology can be improved by increasing the fill factor of the assembled components, optimizing the lamination thickness for reduced eddy current loss, and increasing its power density through shape optimization for achieving uniform magnetization. Due to the limited multi-material printing capacity of current SLM systems, two methods are proposed for eddy current reduction: the interlocking and the gapped core designs. With next-generation powder deposition methods [19], multiple metal or intermetallic materials can be utilized in parallel, allowing for more options and more advanced core topologies.

First, for increasing the fill factor, higher accuracy of the printing system must be achieved. With the current settings, the printed parts still suffer from low surface roughness-related reduced fill factor for interlocking designs or inter-lamination short-circuits and sintered unremovable powder for the gapped designs. Secondly, the lamination thickness can be optimized to provide minimal core losses with maximum part fill factor, i.e., to achieve the optimal ratio of air gap to lamination width. Thirdly, the shape of the core can be optimized for achieving uniform magnetization, weight reduction and improved thermal capacity. Several methods for improving ferromagnetic part performance through topology optimization are discussed in further detail in [20,21]. For improved heat exchange of the printed transformer, enhanced convective heat transfer can easily be obtained by increasing its outer surface area with different surface relief structures [22].

6. Conclusions

In this paper, a fully functioning, additively manufactured soft magnetic transformer core was fabricated and tested. For the first time in literature, an electromagnetic device with a fully 3D printed magnetic core was evaluated in terms of efficiency and performance. The prototype core showed uncompetitive performance when compared to modern conventional transformer cores. Although the printed material is not currently suitable for the production of commercial transformer cores, the analysis of the prototype core did allow us to demonstrate the current state of the art, identify the technical challenges involved, and propose next steps for realizing topology optimized 3D printing soft ferromagnetic components.

A novel, interlocking core design was developed and utilized successfully for achieving a relatively high fill factor of 89% (compared to other 3D printed cores) and eddy current reduction of the additively manufactured transformer core. For obtaining higher fill factor with this method, lower surface roughness of the printed parts must be obtained for more precise fitting of the components. Furthermore, the interlocking core design enabled the integration of modular winding to the transformer design, simplifying its assembly process.

The first prototype transformer core showed both lower efficiency (10% reduced) and power density (34% reduced), when compared to the conventional modern transformer at their respective optimal working conditions. These preliminary performance results of the first prototype core are likely to improve with more refined core designs and materials as part of future research. Currently, the main challenge in realizing high-performance 3D printed soft magnetic components is achieving a higher degree of control over the printed material grain texture, since the conventional post-processing methods for Goss textured silicon steel sheets are not suitable for processing geometrically complex 3D printed magnetic components. Even so, for non-grain-oriented applications (such as rotating electrical machines), the current material properties appear suitable, especially with the unprecedented prototyping freedom of 3D printing systems—which could enable the emergence of entirely new types of machines. Although the current 3D printed cores for AC applications suffer either from high eddy current losses or low filling factor, next-generation emerging multi-metal SLM printers can potentially improve the additively manufactured core performance considerably. Future work on this project will include further optimization of both the printed material and component topology for designing and constructing AM topology optimized electrical machines.

Author Contributions: Conceptualization: A.K. and H.T; methodology: A.R.; validation, A.K.; investigation, H.T. and P.S.G.; resources, P.S.G.; writing—original draft preparation, H.T.; writing—review and editing, H.T. and A.K.; supervision, A.K. and A.B.; project administration, T.V.; funding acquisition, A.K. All authors have read and agreed to the published version of the manuscript.

Funding: This research work has been supported by the Estonian Ministry of Education and Research (Project PSG-137).

Institutional Review Board Statement: Not applicable.

Informed Consent Statement: Not applicable.

Data Availability Statement: Data is contained within the article.

Acknowledgments: The authors would like to thank Balti Transformers Ltd. for cooperation.

Conflicts of Interest: The authors declare no conflict of interests.

References

1. How 3D Printing is Redefining Inductor Coil Production | GKN Additive. Available online: https://www.gknpm.com/en/our-businesses/gkn-additive/how-3d-printing-is-redefining-inductor-coil-production/ (accessed on 26 November 2020).
2. GE Aviation 3D Prints 30,000th Metal 3D Printed Fuel Nozzle at Auburn, Alabama Plant. Available online: https://3dprint.com/226703/ge-aviation-fuel-nozzle-3d-printed-30000/?fbclid=IwAR38NO-0dAf2BwIkXhVtlC18gAimKnQUSy5KRv-u08InfruK5Zl7Ql1P1HI (accessed on 26 November 2020).
3. Conflux Technology is Reinventing Heat Exchangers with 3D Printing » 3D Printing Media Network—The Pulse of the AM Industry. Available online: https://www.3dprintingmedia.network/conflux-technology-reinventing-heat-exchangers/ (accessed on 26 November 2020).
4. Ngo, T.D.; Kashani, A.; Imbalzano, G.; Nguyen, K.T.Q.; Hui, D. Additive manufacturing (3D printing): A review of materials, methods, applications and challenges. *Compos. Part B Eng.* **2018**, *143*, 172–196. [CrossRef]
5. Tiismus, H.; Kallaste, A.; Belahcen, A.; Rassolkin, A.; Vaimann, T. Challenges of Additive Manufacturing of Electrical Machines. In Proceedings of the 2019 IEEE 12th International Symposium on Diagnostics for Electrical Machines, Power Electronics and Drives (SDEMPED), Toulouse, France, 27–30 August 2019; pp. 44–48. [CrossRef]
6. Tiismus, H.; Kallaste, A.; Belahcen, A.; Tarraste, M.; Vaimann, T.; Rassõlkin, A.; Asad, B.; Ghahfarokhi, P.S. AC Magnetic Loss Reduction of SLM Processed Fe-Si for Additive Manufacturing of Electrical Machines. *Energies* **2021**, *14*, 1241. [CrossRef]
7. Stornelli, G.; Faba, A.; Di Schino, A.; Folgarait, P.; Ridolfi, M.R.; Cardelli, E.; Montanari, R. Properties of additively manufactured electric steel powder cores with increased si content. *Materials* **2021**, *14*, 1489. [CrossRef] [PubMed]
8. Yap, C.Y.; Chua, C.K.; Dong, Z.; Liu, Z.H.; Zhang, D.Q.; Loh, L.E.; Sing, S.L. Review of selective laser melting: Materials and applications. *Appl. Phys. Rev.* **2015**, *2*, 041101. [CrossRef]
9. Li, R.; Liu, J.; Shi, Y.; Wang, L.; Jiang, W. Balling behavior of stainless steel and nickel powder during selective laser melting process. *Int. J. Adv. Manuf. Technol.* **2012**, *59*, 1025–1035. [CrossRef]
10. Waasner Magnetic and Technological Properties. Available online: http://www.waasner.de/fileadmin/Assets/PDFs/MaterialCharacteristics_201111.pdf (accessed on 1 March 2021).
11. Waasner Material Characterization. Available online: http://www.waasner.de/fileadmin/Assets/PDFs/MaterialCharacteristics_201111.pdf (accessed on 1 March 2021).
12. Data Sheet Isovac 235-35 A. Available online: https://www.voestalpine.com/division_stahl/content/download/39689/456867/file/DB_isovac_235-35A_E_281015.pdf (accessed on 1 March 2021).
13. Tiismus, H.; Kallaste, A.; Belahcen, A.; Vaimann, T.; Rassõlkin, A.; Lukichev, D. Hysteresis Measurements and Numerical Losses Segregation of Additively Manufactured Silicon Steel for 3D Printing Electrical Machines. *Appl. Sci.* **2020**, *10*, 6515. [CrossRef]
14. Block—PCB Transformers. Available online: http://www.farnell.com/datasheets/1897307.pdf (accessed on 1 March 2021).
15. Block—FL 30/12 Safety Isolating Transformer. Available online: https://www.block.eu/en_US/productversion/fl-3012/ (accessed on 1 March 2021).
16. Brownsburg Electronic Inc. Transformers and Inductors. Available online: http://www.bei.net/PDF/catalogue2004.pdf (accessed on 1 March 2021).
17. Plotkowski, A.; Carver, K.; List, F.; Pries, J.; Li, Z.; Rossy, A.M.; Leonard, D. Design and performance of an additively manufactured high-Si transformer core. *Mater. Des.* **2020**, *194*, 108894. [CrossRef]
18. Suwas, S.; Ray, R.K. *Crystallographic Texture of Materials*; Springer: London, UK, 2014.
19. Aconity Additive Manufacturing. Available online: https://aconity3d.com/ (accessed on 29 March 2021).
20. Andriushchenko, E.; Kallaste, A.; Belahcen, A.; Vaimann, T.; Rassõlkin, A.; Heidari, H.; Tiismus, H. Optimization of a 3D-Printed Permanent Magnet Coupling Using Genetic Algorithm and Taguchi Method. *Electronics* **2021**, *10*, 494. [CrossRef]

21. Orosz, T.; Rassõlkin, A.; Kallaste, A.; Arsénio, P.; Pánek, D.; Kaska, J.; Karban, P. Robust Design Optimization and Emerging Technologies for Electrical Machines: Challenges and Open Problems. *Appl. Sci.* **2020**, *10*, 6653. [CrossRef]
22. Ghahfarokhi, P.S.; Podgornovs, A.; Kallaste, A.; Cardoso, A.J.M.; Belahcen, A.; Vaimann, T.; Tiismus, H.; Asad, B. Opportunities and Challenges of Utilizing Additive Manufacturing Approaches in Thermal Management of Electrical Machines. *IEEE Access* **2021**, *9*, 36368–36381. [CrossRef]

Article

Sliding Mean Value Subtraction-Based DC Drift Correction of B-H Curve for 3D-Printed Magnetic Materials

Bilal Asad [1,2,*], **Hans Tiismus** [1], **Toomas Vaimann** [1], **Anouar Belahcen** [1,2], **Ants Kallaste** [1], **Anton Rassõlkin** [1] and **Payam Shams Ghafarokhi** [3]

[1] Department of Electrical Power Engineering and Mechatronics, Tallinn University of Technology, 19086 Tallinn, Estonia; hans.tiismus@taltech.ee (H.T.); toomas.vaimann@taltech.ee (T.V.); Anouar.Belahcen@aalto.fi (A.B.); ants.kallaste@taltech.ee (A.K.); anton.rassolkin@taltech.ee (A.R.)
[2] Department of Electrical Engineering and Automation, Aalto University, FI-00076 Espoo, Finland
[3] Department of Electrical Machine and Apparatus, Riga Technical University, LV-1658 Riga, Latvia; payam.shams-ghahfarokhi@rtu.lv
[*] Correspondence: bilal.asad@aalto.fi

Abstract: This paper presents an algorithm to remove the DC drift from the *B-H* curve of an additively manufactured soft ferromagnetic material. The removal of DC drift from the magnetization curve is crucial for the accurate estimation of iron losses. The algorithm is based on the sliding mean value subtraction from each cycle of calculated magnetic flux density (*B*) signal. The sliding mean values (SMVs) are calculated using the convolution theorem, where a DC kernel with a length equal to the size of one cycle is convolved with B to recover the drifting signal. The results are based on the toroid measurements made by selective laser melting (SLM)-based 3D printing mechanism. The measurements taken at different flux density values show the effectiveness of the method.

Keywords: additive manufacturing; convolution; infinite impulse response (IIR) filters; additive white noise; DC drift; magnetic flux density; magnetic hysteresis; kernel; magnetic materials

Citation: Asad, B.; Tiismus, H.; Vaimann, T.; Belahcen, A.; Kallaste, A.; Rassõlkin, A.; Ghafarokhi, P.S. Sliding Mean Value Subtraction-Based DC Drift Correction of B-H Curve for 3D-Printed Magnetic Materials. *Energies* **2021**, *14*, 284. https://doi.org/10.3390/en14020284

Received: 3 December 2020
Accepted: 4 January 2021
Published: 6 January 2021

Publisher's Note: MDPI stays neutral with regard to jurisdictional claims in published maps and institutional affiliations.

1. Introduction

Unlike subtractive and injection molding-based manufacturing techniques, additive manufacturing (AM) is gaining heightened popularity. It is also known as 3D printing, where any object can be built layer upon layer using any AM technique. The most common AM techniques include powder bed fusion (PBF), bed jetting (BJ), direct energy deposition (DED), material extrusion (ME), material jetting (MJ), sheet lamination (SL), vat polymerization (VP), etc. However, in electrical machines, direct laser melting (DLM) (a subcategory of PBF) has proven its effectiveness. Various materials such as metals, thermoplastics, ceramics, and biochemicals can be handled through AM techniques. The capability to design complex geometries, to create relatively lightweight products, to perform rapid prototyping, and to save time, as well as the lack of need for fixtures or dies, are the prominent advantages of this technology.

The growing AM-related technological advancements and complex geometries of electrical machines are persuading researchers to test 3D printing in this field. Electrical machines contain various materials such as copper, silicon steel (SiFe), insulation materials, etc. Although the printing of a complete machine in one set has not been achieved so far, the production of separate portions and their assembly afterward is used for various machines. The printing and assembly of motor parts, such as soft magnetic rotors [1,2], stators [3,4], electrical windings [5,6], bearings [7], heat exchangers [8,9], and insulations [5], can lead toward the production of any electrical machine.

Electrical machines are a combination of non-magnetic, with high electric conductivity, magnetic, with low electric conductivity, and dielectric materials, with no electric conductivity. These materials make the coils, cores, and insulation parts of electrical machines. The selection of any appropriate composite material with fixed proportions of different elements is essential for better electrical machine performance. The composition of different materials can have a prominent effect on the core's magnetic properties, which can change the iron losses. The printing of several different materials using selective laser melting (SLM) is available in the literature. The SLM fabrication of conductive (Cu [10], AlSi10MG [11]) and soft ferromagnetic (FeSi6.7 [12], FeSi6.9 [13], FeSi3 [14], Fe-Co-V [15], Fe-Ni-Si [16]) materials are well explained in the literature.

A slight variation in the proportion of the materials can produce dramatic changes in their magnetic characteristics. This makes the investigation of the *B-H* curve of the design material very important. The most common way to characterize any magnetic material is through its hysteresis loop measurement. Commercial hysteresis loop tracers are designed for traditional hard magnetic materials. For soft magnetic materials, custom measurement setups, designed according to the sample's particularities and the measurement objectives, are used. Unlike planar samples, closed circuits (e.g., toroid) are good choices to determine the material's magnetic characteristics. The toroidal structure is preferred because it has the least demagnetizing effects. Furthermore, they can be used to obtain magnetic anisotropy by torque magnetometry.

The biggest challenge in the measurement of an accurate hysteresis loop is the integrator DC drift. The leading causes of this drift include data acquisition setup offset values, thermal shift of parameters, impedance mismatch among various components of the measurement setup, and additive white gaussian noise (AWGN). This drift becomes very significant when many cycles need to be measured, such as toroidal samples where the maximum magnetic flux density is much larger in value than the coercivity [17]. The coercivity and remanence measurement depends upon the points where the hysteresis loop cut the *B* and *H* axis. Hence those points should be in a narrow range. Commercial integrators give a correction range up to 2 μWb/min [18], which is not enough for signals with a long measurement time. Some solutions to remove the DC drift are proposed in the literature, but they possess drawbacks, such as the following:

(1) The constant drift assumption [19–21] does not remain valid when the measurement time is extended. This is due to the presence of switching frequency-based noise in the digital data acquisition setup, which worsens the problem.
(2) The removal of DC drift by subtracting an approximate polynomial data fitting function from the signal [17] does not consider all cycles separately. It can give good results when the number of measured cycles is high enough, which is difficult to obtain under low-frequency measurements—furthermore, the presence of switching noise increases approximation.

This work presents an algorithm to effectively remove the drift due to the DC offset of the data acquisition setup. This technique has not been previously presented in literature to the best of the authors' knowledge. Unlike constant and approximate polynomial data fitting functions, the proposed algorithm detects the drift at each sample, making it independent of signal length. For this purpose, the measured signals are convolved with a kernel function to get a sliding mean value function (SMVF). The subtraction of SMVF from corresponding measured signals reduces the DC drift significantly. Additionally, the high-frequency switching noise is reduced using infinite impulse response (IIR) low pass filters [22,23]. For the validation, a 3D-printed toroid is tested under different values of maximum flux density.

2. The Proposed Algorithm

A flowchart diagram of the proposed algorithm is presented in Figure 1. A detailed explanation of all steps is provided below.

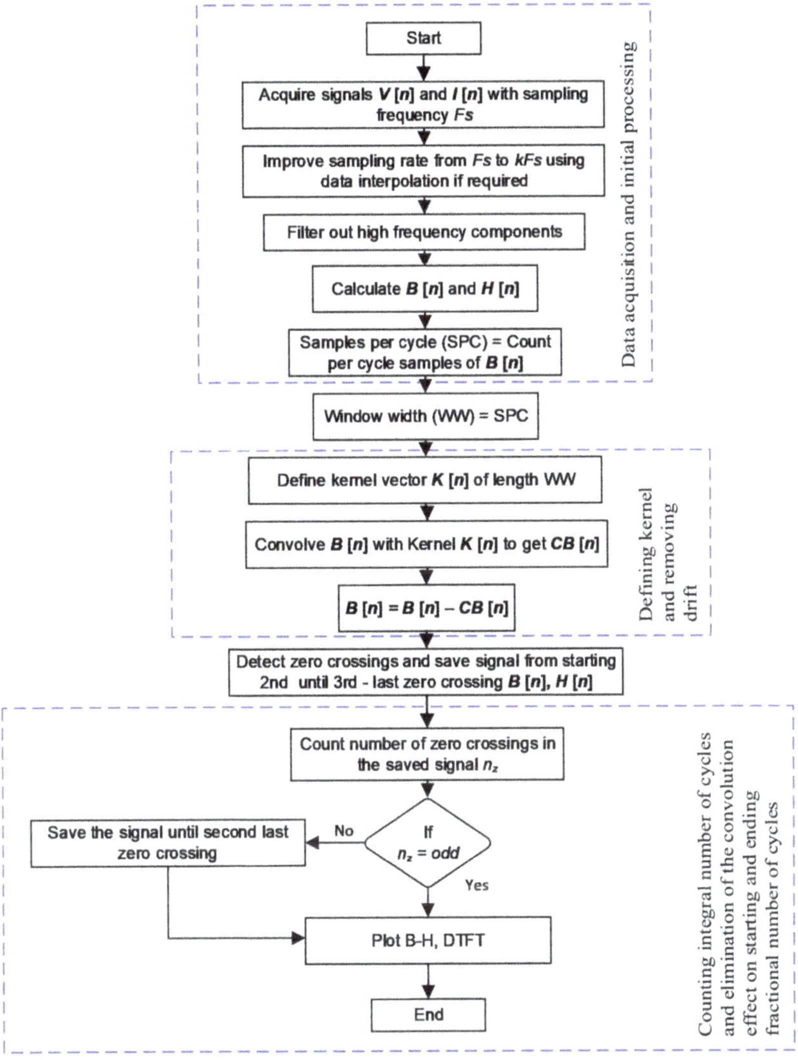

Figure 1. The algorithm flowchart.

2.1. Data Acquisition and Initial Processing

- Measure the input primary current I [n] and output secondary voltage V [n] with reasonable sampling frequency F_s. If spectrum analysis is also required along with the B-H curve, the minimum sampling frequency should follow the Nyquist criterion.
- The sampling frequency can be improved by data interpolation. Linear interpolation to increase sampling frequency from F_s to kF_s can have less impact on the signal's shape. An increase in the sampling frequency is necessary to recover zero crossings in the signal. If the sampling rate is not good enough, the approximate zero-crossings would be away from the actual zero line, as shown in Figure 2. This may lead to signal processing-related problems, such as spectral leakage or increased ripples due to bandpass filters. However, very powerful data acquisition devices are available these days, and data interpolation can make the algorithm suitable for implementation in on-board processors.

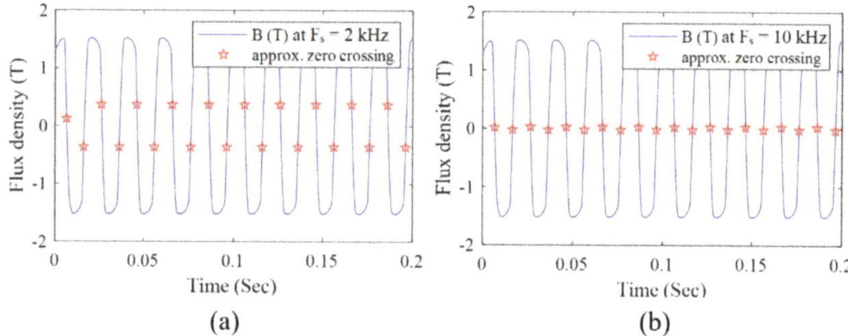

Figure 2. The impact of sampling frequency on zero-crossing detection of flux density *B* measured at 50 Hz: (**a**) at a low sampling frequency (2 kHz), (**b**) at an improved sampling frequency (10 kHz).

- Random noise, such as additive white gaussian noise (AWGN), is inevitable in home-made *B-H* tracers. This is because of the high frequency switching noise if digital signal generators and power supplies are used. Moreover, this noise also occurs because of the mismatched impedances between various components of the experimental setup. Although AWGN noise does not significantly contribute to the DC drift, it causes random movement of the *B* and *H* intercept points, mainly if the signal to noise ratio (SNR) is low. Being random, AWGN noise cannot be eliminated completely; however, its impact can be minimized by attenuating the high-frequency noise components using an IIR low pass filter. IIR filters are the right choice because of their narrow transition band and fewer passband ripples.
- The magnetic field strength (*H(t)*) and magnetic field density (*B(t)*) can be calculated using the following formulas.

$$H(t) = \frac{N_1 i_1(t)}{l} \tag{1}$$

$$B(t) = \frac{1}{N_2 A_c} \int_{t_i}^{t_f} v_2(t) d(t) \tag{2}$$

where N_1 and N_2 are the number of primary and secondary turns, respectively, $i_1(t)$ is the current in the primary winding, l is the average length of the toroid taken at the center, A_c is the core cross-sectional area, $v_2(t)$ is the secondary induced voltage, and t_i and t_f represent the measurement interval and signal length, respectively. It is important to know that the measured induced voltage contains the drift signal acting as a constant of integration.

The number of samples per cycle (SPC) is vital to define the length of the kernel function. For the known values of sampling (F_s) and supply frequency (f_s), the size of the one cycle of *B* or *H* can be calculated as follows:

$$SPC = \frac{F_s}{f_s} \tag{3}$$

2.2. Defining Kernel Function and the Removal of the DC Drift

Convolution is a potent tool in signal processing, where two functions generate a third function that describes how the shape of one signal is modified by the other. In digital signal processing, the convolution of two functions *f* and *g* can be defined as follows:

$$(f * g)[n] = \sum_{m=-M}^{M} f[m]g[n-m] \tag{4}$$

where f is the input vector, which is $B[n]$ in our case, and g is the kernel vector defined by (5), which acts as a filter. The size and shape of the kernel is the most significant part in signal processing. Since our goal is to find out the signal's envelope pattern, a constant kernel vector of size equal to one cycle of the input signal can be the right choice. Furthermore, to eliminate the effect of the kernel itself, the sum of its elements should be equal to unity, as shown below.

$$k[n] = \frac{ones(SPC, 1)}{SPC} \tag{5}$$

The convolution will give the moving mean value function across the entire input signal by making the kernel size equal to one cycle's length. A detailed description of the proposed convolution is presented in Figure 3. Figure 4 shows the rising trend of the envelope of B, which is recovered by convolving the flux density function ($B[n]$) with the proposed kernel function $k[n]$. The trend curve is brought to the signal's surface for ease of understanding, although it remains in the center of the signal as a mean value function. Subtracting the recovered mean value function from the original signal (B) removes the DC drift.

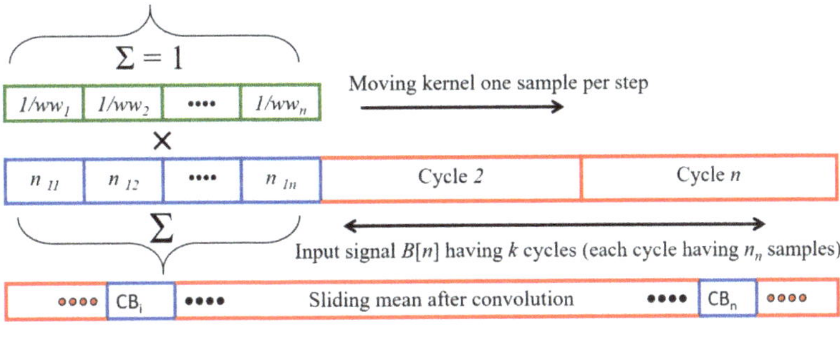

$$CB_i = \frac{n_{11}}{ww_1} + \frac{n_{12}}{ww_2} + \cdots\cdots + \frac{n_{1n}}{ww_n}, \qquad CB_{i+1} = \frac{n_{12}}{ww_1} + \frac{n_{13}}{ww_2} + \cdots\cdots + \frac{n_{21}}{ww_n}$$

$$CB_n = \frac{n_{n1}}{ww_1} + \frac{n_{n2}}{ww_2} + \cdots\cdots + \frac{n_{nn}}{ww_n}. \qquad n_{\,i(cycle)\,j(sample)}$$

Figure 3. Description of the moving mean value technique.

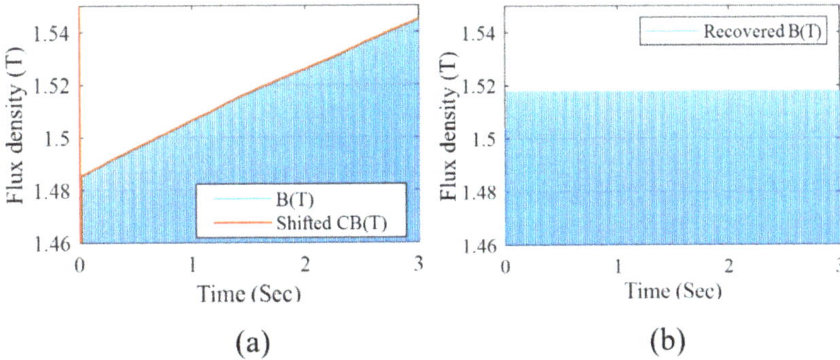

Figure 4. The envelope of the flux density $B(t)$ (**a**) with DC drift, shown by brown line recovered by the proposed algorithm; (**b**) the corrected signal after removing DC drift.

It is essential to know that when the kernel's overlapping size is not equal to the signal's cycle length, the non-overlapping elements of kernel functions are considered zero during the starting and ending intervals. This leads to the transient interval during the resultant signal's starting and ending points, represented by red dots in Figure 3. Since the length of those intervals is equal to one cycle's size, the mean value signal should be saved from CB_i to CB_n. The initial and last cycles can be discarded using zero-crossing detection.

2.3. Zero-Crossing Detection

Almost all filters have transient and steady-state intervals. The transient interval appears at the beginning and the end of the signal. The duration of the transient interval depends upon the type of the filter and the windowing function. Since in the proposed algorithm, two filters, lowpass IIR and convolution, are used, the transient interval is very narrow. The resultant signal can be saved from the second zero crossing to the third last zero crossings to achieve the steady-state interval. The resultant acquired signal in the steady-state interval can have an integral or fractional number of cycles. Although the number of cycles does not significantly impact the *B-H* curve, they can affect the frequency resolution of fast Fourier transform (FFT) or other spectrum analysis-based signal processing techniques. Since each sinusoidal signal has three zero crossings, the integral number of cycles would have an odd number of zero crossings. This fact is used to save the signal in such a way that if the number of zero crossings is odd, then the signal is ready for further analysis; otherwise, the signal can be saved from first zero crossings until second last. Due to the ease in detecting the starting and ending fractional cycles, the method of zero-crossing detection for counting the integral number of cycles is preferred over the conventional approach where sampling frequency and the measurement length of the signal can be used for the same purpose.

3. Measurement Setup

The experimental setup consists of two parts: printing and measurement. The sample toroid is prepared from Fe-Si powder with powder bed fusion printer Realizer SLM-50 (Germany). Laser re-melting strategy (each printed layer is scanned twice before applying the next layer of powder) is utilized for reducing the irregularities in the solidified part. The laser beam power was chosen as 50 W (1 m/s) for the primary and 75 W (0.75 m/s) for the secondary scan. The printed toroidal sample exhibited a 5 mm × 5 mm rectangular cross-section and a 60 mm outer diameter. The toroid was post-processed in a vacuum furnace for internal stress normalization and structure recrystallization. The toroid was heated 300 K/h up to the temperature of 1150 °C and was annealed in a vacuum chamber at 1150 °C for one hour and then furnace cooled. Q detailed description of the printing procedure is presented in [24,25].

The printed toroid was wound with primary (inner, magnetizing) and secondary coils (outer, measuring) with 150 (N_1) and 50 (N_2) turns. All the measurements were conducted per European standards EN 60404-4 and EN 60404-6 [26,27]. The primary coil was supplied with a sinusoidal current with different peak values to maintain flux density in the range of 0.5–1.6 T. The sinusoidal current was generated using an Omicron power amplifier CMS 356 (Austria), which was fed with a sinusoidal reference signal coming from a digital function generator, as shown in Figure 5. The current was measured using a precision resistor 75 mV/15 A connected in series with the primary winding. This current calculates magnetic field strength ($H(t)$) as in (1), while magnetic flux density ($B(t)$) is calculated from the output voltage across the secondary winding using (2). All these measurements were taken with a considerably high sampling frequency of 10 kHz using data acquisition setup Dewetron DEWE2-M (Austria). However, the measurements can be artificially improved using data interpolation, as discussed earlier.

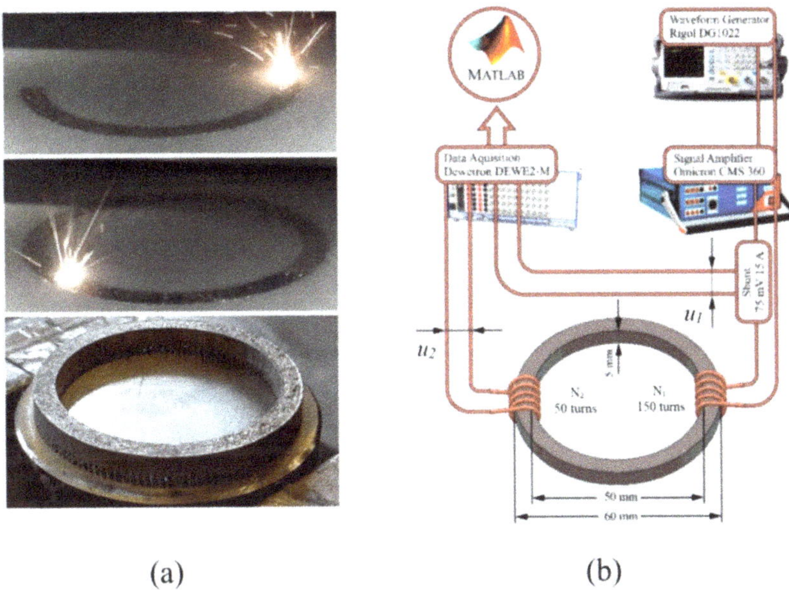

(a)   (b)

Figure 5. Experimental setup of (**a**) the printing of the sample and (**b**) the measurement schematic diagram.

4. Results and Discussion

The measured hysteresis loop with different maximum flux densities (B_{max}) is shown in Figure 6. Figure 6a shows the results without any significant signal processing. As discussed earlier, the flux density and field strength vectors are calculated from the measured voltage and currents. Only the B vector is normalized across the zero line by subtracting it with its mean value. This is important; otherwise, different loops will have other locations due to significant DC shifts. The B and H intercept drift is visible by the lines' thickness as the drift due to DC offset is still there. Figure 6b shows the recovered loops after removing DC drift using 3rd-degree polynomial function. The polynomial function creates an approximate fitting function at the center of the vector. Hence, the drift can be removed by subtracting the $B(t)$ from the polynomial. The drift is somewhat reduced, but still, the line widths are considerable. This is because of the approximation considered by the polynomial function. This problem worsens if the number of measurement cycles is less in number. Figure 6c shows the recovered hysteresis loops using the proposed algorithm. As compared to Figure 6a,b, the drift is considerably reduced in Figure 6c. Moreover, the proposed algorithm does not depend upon the number of cycles under consideration because it considers every cycle independently.

Figure 7 compares recovered loops using the conventional polynomial fitting function with ones obtained using the proposed algorithm. The zoomed windows in Figure 7a,b depict the B-H curve's sharpening using the proposed algorithm.

The approximate errors using different techniques are given in Table 1. This error is calculated by subtracting the higher (B^{+h}) and lower intercept (B^{+l}) values on the B axis at zero H. The error is considerably reduced from 0.05 T (intercept width) to 0.0002 T using the proposed algorithm.

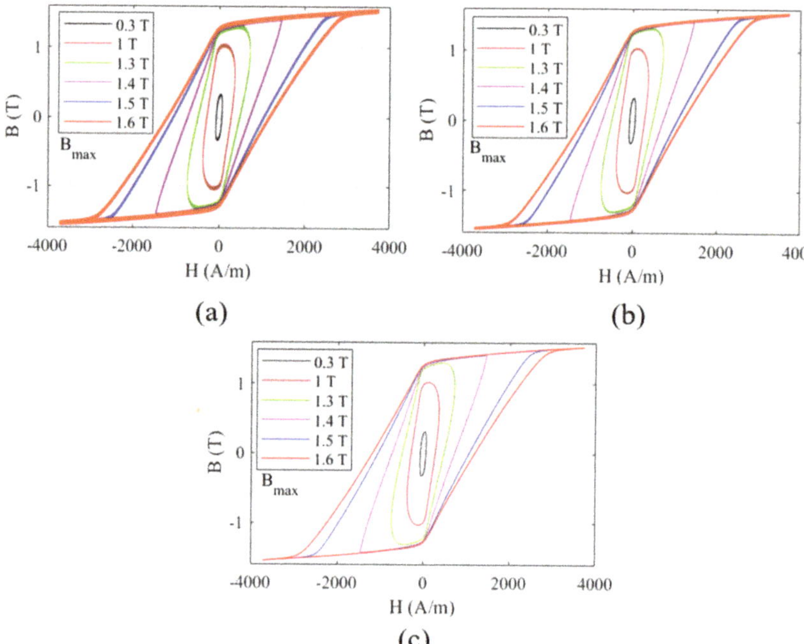

Figure 6. The measured *B-H* curves at different maximum flux densities B_{max} (**a**) without DC drift correction; (**b**) DC drift correction using the polynomial fitting function as in [17]; (**c**) the corrected *B-H* curves using the proposed algorithm.

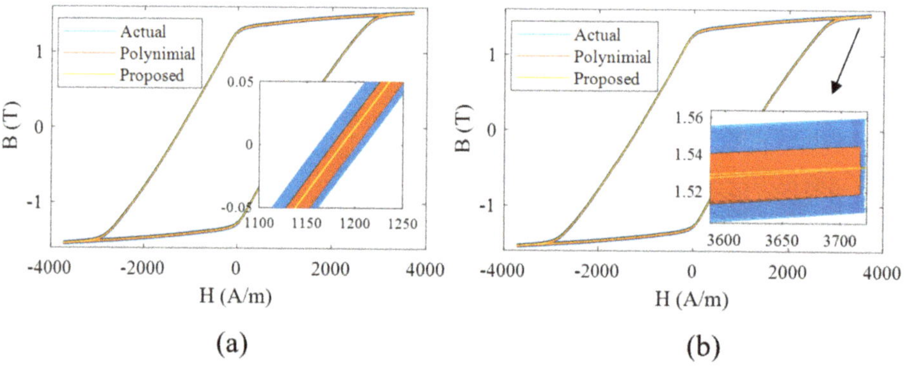

Figure 7. A comparison of the different approaches, (**a**) with zoomed window near zero crossing of the *B* axis, (**b**) with zoomed window at the peak value.

Table 1. The comparison of different drift removal techniques in terms of *B* intercept width.

Sr. No.	Technique	B-Axis Intercept Range at $H = 0$ (Error) $\Delta B = B^{+h} - B^{+l}$
1	Measured	0.05 T (5%)
2	Polynomial adjustment based	0.02 T (2%)
3	Proposed	0.0002 T (0.02%)

5. Conclusions

Advancements in SLM-based AM techniques have opened a broad domain of intricate electrical machine designs. This complexity can be in the form of optimized complex mechanical geometry or the selection of different types of composite material for efficiency improvement. The selection of composite material to be used for fabrication purposes depends upon its magnetic properties, which directly influence efficiency. Hence, before the final selection of machine-fabricated material, an evaluation of its characteristics is mandatory. The percentage content of different elements in the fabricating power can be optimized based on evaluations of those characteristic. In electrical machines, the material's magnetic characteristics are the most crucial parameter among several others. For this purpose, an algorithm was proposed in this paper, which was shown to produce the *B-H* curve of the 3D printed sample with less error. The proposed model was shown to have the following benefits:

- The proposed algorithm can work on the least number of measured current or voltage cycles compared to the corresponding data fitting-based correction algorithm.
- The higher-order noise components because of switching frequency, AWGN, and the impedance mismatch are effectively removed using the IIR low-pass filter. However, it can be avoided if the signal has a high signal to noise ratio (SNR).
- The increase in the sampling rate using data interpolation makes the algorithm suitable to plot good magnetic characteristics from the data measured at a low sampling frequency. This makes algorithms convenient, even if the data acquisition devices are not very powerful.
- The zero-crossing detection helps to count the integral number of cycles and removes the filter-based transient intervals. The removal of the transient interval is very crucial to getting a smooth *B-H* curve. Additionally, it reduces the spectral leakage if the frequency spectrum analysis is required.
- The proposed algorithm gives a very narrow band intercept across the *B* and *H* axis, which reduces the error considerably.

Author Contributions: Conceptualization: B.A., H.T., T.V., and A.B.; methodology: B.A., H.T., T.V., and A.B.; validation: A.K., A.R., and A.B.; data curation: B.A. and H.T.; writing—original draft preparation: B.A.; writing—review and editing: B.A., H.T., and A.R.; visualization: T.V. and P.S.G.; supervision: T.V., A.K., and A.B. All authors have read and agreed to the published version of the manuscript.

Funding: This research work has been supported by Estonian Ministry of Education and Research (Project PSG137).

Conflicts of Interest: The authors declare no conflict of interest.

References

1. Lammers, S.; Adam, G.; Schmid, H.J.; Mrozek, R.; Oberacker, R.; Hoffmann, M.J.; Quattrone, F.; Ponick, B. Additive Manufacturing of a lightweight rotor for a permanent magnet synchronous machine. In Proceedings of the 2016 6th International Electric Drives Production Conference, EDPC 2016-Proceedings, Nuremberg, Germany, 30 November–1 December 2016; pp. 41–45. [CrossRef]
2. Garibaldi, M.; Gerada, C.; Ashcroft, I.; Hague, R. Free-Form Design of Electrical Machine Rotor Cores for Production Using Additive Manufacturing. *J. Mech. Des.* **2019**, *141*, 1–27. [CrossRef]
3. Metsä-Kortelainen, S.; Lindroos, T.; Savolainen, M.; Jokinen, A.; Revuelta, A.; Pasanen, A.; Ruusuvuori, K.; Pippuri, J. Manufacturing of topology optimized soft magnetic core through 3D printing. In Proceedings of the NAFEMS Exploring the Design Freedom of Additive Manufacturing through Simulation, Helsinki, Finland, 22–23 November 2016.
4. Garibaldi, M. Laser additive Manufacturing of Soft Magnetic Cores for rotating Electrical Machinery: Materials Development and Part Design. Ph.D. Thesis, University of Nottingham, Nottingham, UK, 2018.
5. Lorenz, F.; Rudolph, J.; Wemer, R. Design of 3D Printed High Performance Windings for Switched Reluctance Machines. In Proceedings of the 23rd International Conference on Electrical Machines, ICEM 2018, Alexandroupoli, Greece, 3–6 September 2018; pp. 2451–2457. [CrossRef]
6. Simpson, N.; Mellor, P.H. Additive manufacturing of shaped profile windings for minimal AC loss in gapped inductors. In Proceedings of the 2017 IEEE International Electric Machines and Drives Conference (IEMDC), Miami, FL, USA, 21–24 May 2017; pp. 1–7. [CrossRef]

7. Richardson, B.; Love, L.; Tate, J.G. Additive Manufacturing for Low Volume Bearings. 2017. Available online: http://www.osti.gov/scitech/ (accessed on 24 November 2020).

8. Wrobel, R.; Hussein, A. Design Considerations of Heat Guides Fabricated Using Additive Manufacturing for Enhanced Heat Transfer in Electrical Machines. In Proceedings of the 2018 IEEE Energy Conversion Congress and Exposition, ECCE 2018, Portland, OR, USA, 23–27 September 2018; pp. 6506–6513. [CrossRef]

9. Sixel, W.; Liu, M.; Nellis, G.; Sarlioglu, B. Cooling of Windings in Electric Machines via 3D Printed Heat Exchanger. In Proceedings of the 2018 IEEE Energy Conversion Congress and Exposition, ECCE 2018, Portland, OR, USA, 23–27 September 2018; pp. 229–235. [CrossRef]

10. Silbernagel, C.; Gargalis, L.; Ashcroft, I.; Hague, R.; Galea, M.; Dickens, P. Electrical resistivity of pure copper processed by medium-powered laser powder bed fusion additive manufacturing for use in electromagnetic applications. *Addit. Manuf.* **2019**, *29*, 100831. [CrossRef]

11. Silbernagel, C.; Ashcroft, I.; Dickens, P.; Galea, M. Electrical resistivity of additively manufactured AlSi10Mg for use in electric motors. *Addit. Manuf.* **2018**, *21*, 395–403. [CrossRef]

12. Goll, D.; Schuller, D.; Martinek, G.; Kunert, T.; Schurr, J.; Sinz, C.; Schubert, T.; Bernthaler, T.; Riegel, H.; Schneider, G. Additive manufacturing of soft magnetic materials and components. *Addit. Manuf.* **2019**, *27*, 428–439. [CrossRef]

13. Garibaldi, M.; Ashcroft, I.; Lemke, J.; Simonelli, M.; Hague, R. Effect of annealing on the microstructure and magnetic properties of soft magnetic Fe-Si produced via laser additive manufacturing. *Scr. Mater.* **2018**, *142*, 121–125. [CrossRef]

14. Plotkowski, A.; Pries, J.; List, F.; Nandwana, P.; Stump, B.; Carver, K.; Dehoff, R. Influence of scan pattern and geometry on the microstructure and soft-magnetic performance of additively manufactured Fe-Si. *Addit. Manuf.* **2019**, *29*, 100781. [CrossRef]

15. Riipinen, T.; Metsä-Kortelainen, S.; Lindroos, T.; Keränen, J.S.; Manninen, A.; Pippuri-Mäkeläinen, J. Properties of soft magnetic Fe-Co-V alloy produced by laser powder bed fusion. *Rapid Prototyp. J.* **2019**, *25*, 699–707. [CrossRef]

16. Kang, N.; El Mansori, M.; Guittonneau, F.; Liao, H.; Fu, Y.; Aubry, E. Controllable mesostructure, magnetic properties of soft magnetic Fe-Ni-Si by using selective laser melting from nickel coated high silicon steel powder. *Appl. Surf. Sci.* **2018**, *455*, 736–741. [CrossRef]

17. Garcia, J.; Rivas, M. A quasi-static magnetic hysteresis loop measurement system with drift correction. *IEEE Trans. Magn.* **2006**, *42*, 15–17. [CrossRef]

18. Horn, J.L.; Grimes, C.A. A 0.1–500 Hz analog thin film BH hysteresis loop tracer with automatic Y-axis drift correction. *Rev. Sci. Instrum.* **1997**, *68*, 1346–1347. [CrossRef]

19. Franco, V.; Ramos-Martos, J.; Conde, A. Autocalibrating quasistatic M-H hysteresis loop tracer with negligible drift. *Rev. Sci. Instrum.* **1996**, *67*, 4167–4170. [CrossRef]

20. Krings, A.; Soulard, J.; Wallmark, O. PWM Influence on the Iron Losses and Characteristics of a Slotless Permanent-Magnet Motor with SiFe and NiFe Stator Cores. *IEEE Trans. Ind. Appl.* **2015**, *51*, 1475–1484. [CrossRef]

21. Krings, A.; Nategh, S.; Wallmark, O.; Soulard, J. Influence of the Welding Process on the Performance of Slotless PM Motors With SiFe and NiFe Stator Laminations. *IEEE Trans. Ind. Appl.* **2014**, *50*, 296–306. [CrossRef]

22. Asad, B.; Vaimann, T.; Belahcen, A.; Kallaste, A.; Rassõlkin, A.; Iqbal, M.N. Broken rotor bar fault detection of the grid and inverter-fed induction motor by effective attenuation of the fundamental component. *IET Electr. Power Appl.* **2019**, *13*, 2005–2014. [CrossRef]

23. Asad, B.; Vaimann, T.; Kallaste, A.; Rassõlkin, A.; Belahcen, A.; Iqbal, M.N. Improving Legibility of Motor Current Spectrum for Broken Rotor Bars Fault Diagnostics. *Electr. Control. Commun. Eng.* **2019**, *15*, 1–8. [CrossRef]

24. Tiismus, H.; Kallaste, A.; Belahcen, A.; Vaimann, T.; Rassõlkin, A.; Lukichev, D.V. Hysteresis measurements and numerical losses segregation of additively manufactured silicon steel for 3d printing electrical machines. *Appl. Sci.* **2020**, *10*, 6515. [CrossRef]

25. Tiismus, H.; Kallaste, A.; Belahcen, A.; Rassolkin, A.; Vaimann, T. Hysteresis Loss Evaluation of Additively Manufactured Soft Magnetic Core. In Proceedings of the 2020 International Conference on Electrical Machines (ICEM), Kärdla, Estonia, 12–15 June 2019; pp. 1657–1661. [CrossRef]

26. EN IEC 60404-6:2018-Magnetic Materials-Part 6: Methods of Measurement of the Magnetic Properties of Magnetically Soft Metallic and Powder Materials at Frequencies in the Range 20 Hz to 100 kHz by the Use of Ring Specimens. Available online: https://standards.iteh.ai/catalog/standards/clc/0f265b9d-6454-4597-aa62-9290d0e03700/en-iec-60404-6-2018 (accessed on 27 November 2020).

27. EN 60404-4:1997-Magnetic Materials-Part 4: Methods of Measurement of d.c. Magnetic Properties of Magnetically Soft Materials. Available online: https://standards.iteh.ai/catalog/standards/clc/a53d9e5b-537f-4e4d-87d2-a4fceadb1534/en-60404-4-1997 (accessed on 27 November 2020).

Article

Reducing Rotor Temperature Rise in Concentrated Winding Motor by Using Magnetic Powder Mixed Resin Ring

Mitsuhide Sato [1],*, Keigo Takazawa [1], Manabu Horiuchi [1], Ryoken Masuda [1], Ryo Yoshida [1], Masami Nirei [2], Yinggang Bu [1] and Tsutomu Mizuno [1]

[1] Faculty of Engineering, Shinshu University, Matsumoto 380-8553, Japan; k_takazawa@shinshu-u.ac.jp (K.T.); 19hs207a@shinshu-u.ac.jp (M.H.); 20w2088a@shinshu-u.ac.jp (R.M.); 17t2174j@shinshu-u.ac.jp (R.Y.); buyinggang@shinshu-u.ac.jp (Y.B.); mizunot@shinshu-u.ac.jp (T.M.)

[2] National Institute of Technology, Nagano College, Nagano 381-8550, Japan; m_nirei@nagano-nct.ac.jp

* Correspondence: mitsuhide@shinshu-u.ac.jp; Tel.: +81-26-269-5211

Received: 27 November 2020; Accepted: 17 December 2020; Published: 20 December 2020

Abstract: The demand for high-speed servomotors is increasing, and minimal losses in both high-speed and high-torque regions are required. Copper loss reduction in permanent magnet motors can be achieved by configuring concentrated winding, but there are more spatial harmonics compared with distributed winding. At high-speed rotation, the eddy current loss of the rotor increases, and efficiency tends to decrease. Therefore, we propose a motor in which a composite ring made from resin material mixed with magnetic powder is mounted on the stator to suppress spatial harmonics. This paper describes three characteristic motor types, namely, open-slot motors, composite-ring motors, and closed-slot motors. Spatial harmonics are reduced significantly in composite-ring motors, and rotor eddy current loss is reduced by more than 50% compared with open-slot motors. Thermal analysis suggests that the saturation temperature rise value is reduced by more than 30 K. The use of a composite ring is effective in reducing magnet eddy current loss during high-speed rotation. Conversely, the torque characteristics in the closed-slot motor are greatly reduced as well as the efficiency. Magnetic circuits and simulations show that on electrical steel sheets with high relative permeability, the ring significantly reduces the torque flux passing through the stator, thus reducing the torque constant. To achieve reduced eddy current loss during high-speed rotation while ensuring torque characteristics with the composite ring, it is necessary to set the relative permeability and thickness of the composite ring according to motor specifications.

Keywords: eddy current loss; heat generation; magnetic circuit; magnetic composite material; spatial harmonics; concentrated winding motor

1. Introduction

Increasing motor efficiency is required to help prevent global warming [1,2], since more than 50% of the world's total power consumption is caused by motor driving [3], more than 60% of which can be attributed to the industrial sector [4]. Electrification is progressing in automobiles and aircraft, and it is expected that the demand for increased motor efficiency will further increase [5–7]. Permanent magnet servomotors capable of accurate positioning and speed control are widely used in the industrial sector [8,9], and the demand for increased motor speed is increasing [10–13]. Therefore, the servomotor is required to reduce the losses associated with the high-speed rotation range and the high-torque range.

Permanent magnet motors have concentrated and distributed winding, and each winding method has its own advantages for improving efficiency. For concentrated winding, the winding resistance is reduced as well as the copper loss, since its coil end can be manufactured shorter than that of the distributed winding [14]. Furthermore, the eddy current loss of the rotor tends to increase at high-speed rotation in the concentrated winding because it has more spatial harmonics than distributed winding [15,16]. Slot combination optimization [17,18], skew [19–21], and closed slots [22,23] are effective in suppressing spatial harmonics. Skew tends to complicate the motor configuration, and the closed slot of the magnetic steel sheet may significantly reduce torque. Using bonded magnets makes it possible to reduce the eddy current loss of the rotor during high-speed rotation [24], but the torque constant tends to decrease compared with sintered rare-earth magnets [25]. A decrease in the torque constant significantly increases copper loss for a large torque. The magnet eddy current loss increases at high-speed rotation, and the temperature rise is high in the concentrated winding servomotor [26]. The bonded magnet motors and closed-slot motors made of electrical steel sheets can reduce magnet eddy current loss, but cannot maintain torque characteristics. For a concentrated winding high-speed servomotor, it is necessary to reduce the eddy current loss of the magnet while maintaining the torque constant.

This paper proposes a method using a resin material [27] mixed with magnetic powder as a method for reducing the spatial harmonics of a concentrated winding motor. The stator is made of magnetic resin in order to reduce the iron loss of the stator [28–30]. The magnetic composite is used for the teeth of the stator core. The advantages of composites have been reported to include their capability to form three-dimensional shapes and the fact that their core loss under high frequency conditions is relatively low [31]. In this paper, spatial harmonics are suppressed by constructing the ring from composite resin and attaching it to the stator. The magnetic composite resin has a relative permeability < 10–100, and it is effective in quelling the suppression of spatial harmonics while preventing the magnetic flux from short circuiting. In this paper, the effect of reducing both the eddy current loss of the rotor and the temperature rise value is clarified by simulation. Furthermore, the required thickness and relative permeability of the composite ring with respect to preventing the magnetic flux from short circuiting while maintaining torque characteristics are outlined with simulation results and magnetic circuit theory.

2. Magnetic Composite Ring Motor

2.1. Magnet Eddy Current Loss

Figure 1a explains the cause of spatial harmonics in the cross-sectional configuration of the open-slot concentrated winding motor. The stator consists of a yoke and winding, and the rotor consists of a permanent magnet and yoke. The stator yoke is laminated with electromagnetic steel sheets punched into the same shape to reduce iron loss. In the direction of the solid arrow in Figure 1a, the magnetoresistance is relatively small due to the presence of the stator core. Conversely, the magnetic resistance to the direction of the dotted arrow in Figure 1a is relatively large due to the composition of copper and air. The magnetic resistance to the rotation angle of the rotor has a rectangular wave shape, and the fluctuation of the magnetic flux passing through the stator is large, so that the spatial harmonic becomes large. In particular, the distributed winding approaches a sinusoidal magnetic flux distribution, but the concentrated winding has a rectangular wave-shaped magnetic flux distribution, so spatial harmonics are likely to occur.

Rare-earth sintered magnets with a high-energy product are often used to increase the power density of motors. However, Nd rare-earth magnets, containing a large amount of iron, have high electrical conductivity, and fluctuating magnetic fields tend to cause eddy current loss [32,33]. In particular, the slot harmonics tend to increase the eddy current loss of the permanent magnet, since the concentrated winding motor has a large stator slot pitch. The carrier harmonics of the inverter also affect the eddy current loss of the magnet, but most of the components are caused by the slot harmonics [34]. The temperature rise of the magnet on the small motor tends to be high because the heat dissipation area of the magnet is sufficient. Eddy currents are proportional to the square of the frequency of the fluctuating magnetic field. The higher the rotation speed of the motor, the higher the demand for reducing the eddy current loss of the magnet. This is because the frequency for the change in magnetic flux applied to the iron core and magnet increases to a high level.

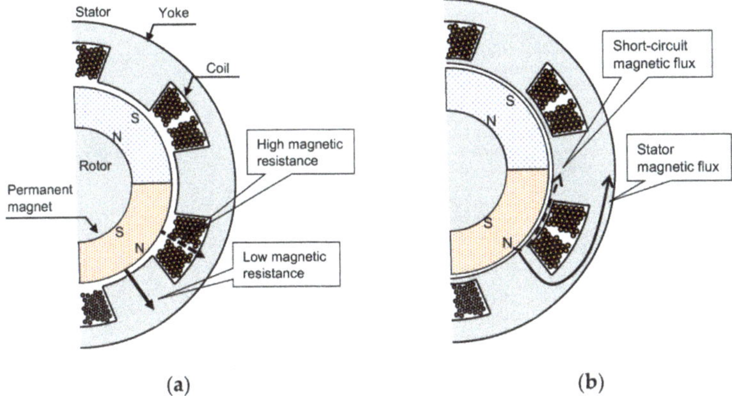

Figure 1. Difference of magnetic flux with open- and closed-slot motor: (**a**) open-slot motor and (**b**) closed-slot motor with an electrical steel sheet. Spatial harmonics are likely to occur in open-slot motors with a large difference in magnetic resistance.

As shown in Figure 1b, the method of forming a closed slot with an electromagnetic steel plate is effective in suppressing spatial harmonics. Since the difference in magnetic resistance between the teeth portion and the coil portion is small when applying the said method, spatial harmonics are suppressed, and the effect of reducing eddy current loss is realized. However, the magnetic flux does not flow to the back yoke, as shown by the dotted arrow in Figure 1b, and the magnetic flux that short-circuits the closed-slot portion increases since the magnetic steel sheet has a high relative magnetic permeability. As shown by the solid arrow in Figure 1b, the torque constant decreases as the stator magnetic flux, which contributes to the generation of torque decreases. Copper loss increases because a current is required when a large torque is generated. Furthermore, when the magnetic flux flows through the closed slot, the magnetic flux density increased, which, in turn, increases iron loss.

Bonded magnets, in which magnets are kneaded into resin, are also effective in reducing magnet eddy current loss. However, they also reduce the gap magnetic flux density. Accordingly, the torque constant decreases, and copper loss tends to increase when a large torque is generated. This is unwanted, as a reduction in magnet eddy current loss is required without significantly reducing the torque constant.

2.2. Structure

Figure 2 shows the structure of a composite-ring motor (hereinafter referred to as a ring motor). Tables 1 and 2 show the structural and material specifications of the ring motor, respectively.

he slot combination consists of four poles and six slots. The outer diameter of the stator is 32 mm, and the shaft length is 40 mm considering the miniaturization as servomotor applications. A ring is provided on the stator. The output of the motor is 500 W.

The difference between a ring motor and a closed-slot motor is that the ring is made of a magnetic composite material instead of electrical steel sheeting. The composite ring formed of a resin material mixed with magnetic powder was mounted on a stator, and windings were embedded with a magnetic material in the ring motor. The magnetic composite material was manufactured by mixing magnetic powder and resin [35,36]. Magnetic composite materials have a higher relative permeability than air, and their relative permeability is significantly smaller than that of electrical steel sheets. Accordingly, the difference in magnetic resistance between the teeth and coil section is reduced, and spatial harmonics can be reduced without suppressing the short circuiting of the magnetic flux. In addition, the magnetic composite material has a high electrical resistance due to the resin. Therefore, it contributes to the reduction in eddy current loss compared to the construction of electrical steel sheets.

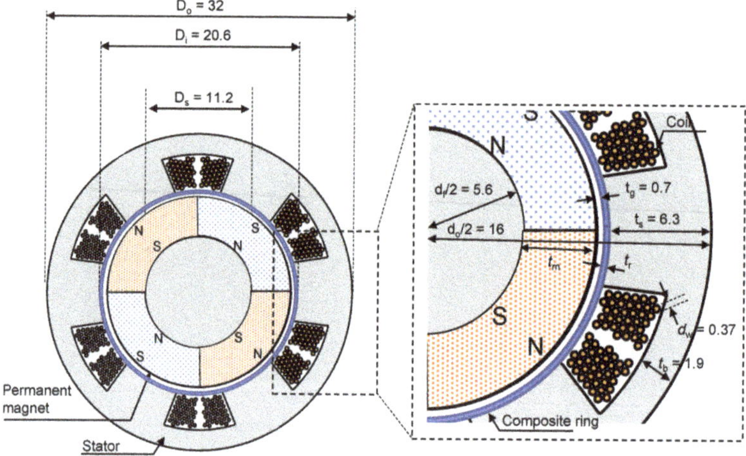

Figure 2. Structure of magnetic ring motor. A ring is provided on the outer circumference of the stator.

Table 1. Motor specifications.

Item	Value
Stator outer diameter D_o	32 mm
Stator inner diameter D_i	20.6 mm
Shaft diameter D_f	11.2 mm
Axial length of motor w_m	40 mm
Number of turns N	10
Gap length t_g	0.7 mm
Wire diameter d_w	0.37 mm
Stator back yoke thickness t_b	1.9 mm
Armature resistance R_a	102 Ω

Table 2. Material specifications.

Item	Value
Stator core	10JNHF600 (JFE Steel Corporation)
Magnetic ring	Amorphous2.6μm + V-2000 64 vol.%
Permanent magnet	NMX-K35CR
Rotor core	35H230

2.3. Magntic Ring Characteristics

The magnetic composite material was manufactured by mixing, stirring, and firing iron amorphous ball powder with an average particle size of 2.6 μm and the impregnated adhesive V-2000 at a volume ratio of 64 vol.% [36]. The magnetic properties were evaluated with a toroidal core made of a magnetic composite material. Figure 3a,b show the measurement results of the magnetization characteristics using a vibrating sample magnetometer (BHV-55: RIKEN-DENSHI) and iron loss characteristics using a B-H analyzer (SY-8218: IWATSU ELECTRIC), respectively. The real part (μ′) of the complex specific magnetic coefficient is 10.4, which is slightly larger than that of air and significantly lower than that of electrical steel sheets. These measurement results will be used for simulations in Chapter 3.

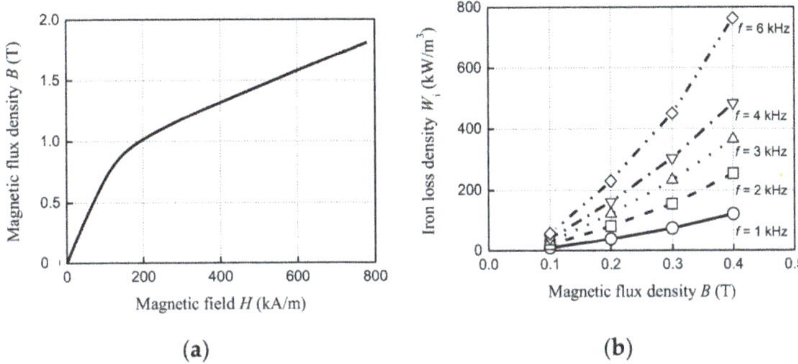

Figure 3. Magnetic characteristics of magnetic composite ring: (**a**) static magnetization and (**b**) iron loss.

3. Effect on Reducing Rotor Temperature Rise

This section clarifies the reducing temperature rise effect of the magnetic ring using simulations. The effect of the magnetic ring motor is confirmed based on the loss of the magnetic field analysis and the temperature rise value of the thermal analysis.

3.1. Simulation Conditions

Table 3 shows the analysis conditions. By comparing the ring motor with three types of open-slot motors without magnetic rings and three types of closed-slot motors with electrical steel sheets, the spatial harmonics and loss suppression effect of the ring motor were confirmed based on the magnetic field analysis. The magnetic field was analyzed using the two-dimensional finite element method (FEM) with JMAG-Designer [37]. The external dimensions of all motors are the same at 32 mm. The gap length was set to 0.7 mm considering winding a synthetic twisted yarn to prevent the magnet from scattering during high-speed rotation. The 0.7 mm gap of the open-slot motor was secured by changing the length of the teeth without changing the outer shape. The armature current was adjusted to ensure the motor output was maintained at 500 W.

Table 3. Analysis conditions.

Item	Value
Software	JMAG-Designer (x64) Ver.18.0
Analysis method	Two-dimensional magnetic field analysis
Solution	FEM
Mesh size	Copper:1/10 or less of the skin depth Ring: Automatic Air: Automatic
Number of mesh	58,938
Mesh type	Slide mesh
Mesh nodes number	29,980
Boundary conditions	Symmetrical boundary
Analysis area	Analysis in 10 times the analysis model
Rotor Speed	$N = 30,000$ rpm
Output	$P_o = 500$ W
Material	Copper: $\rho = 1.72 \times 10^{-8}$ Ωm, $\mu' = 1$, $\mu'' = 0$ Ring: B-H curve and Iron loss profile in Figure 3a,b Air: $\rho = \infty$ Ωm, $\mu' = 1$, $\mu'' = 0$

Thermal analysis was based on the finite element method and the heat conduction equation of JMAG-Designer. Table 4 summarizes the thermal characteristic parameters. The temperature rise was calculated based on the loss of each part derived by magnetic field analysis. Figure 4 shows the thermal circuit assumed in the simulation. The stator consists of a core and coil, and the rotor consists of a core, a magnet, and a shaft. The coil and magnet are the main heat sources. Heat is transferred by contact thermal resistance and dissipated from the rotor surface, the stator core, and the coil into the air by heat transfer. By calculating the temperature rise of each part, the effectiveness of the ring motor in reducing the temperature rise can be clarified.

Table 4. Thermal characteristic parameters.

Item	Value
Software	JMAG-Designer (x64) Ver.18.0
Analysis method	Transient thermal analysis
Thermal conductivity (W/K·m)	Shaft 46.6, Rotor core 46.6, Permanent magnet 7.6, Coil 400, Stator core 18.9, Magnetic ring 3 Between coil and stator core 0.15 (Thickness 0.3 mm) Between magnet and rotor core 0.027 (Thickness 0.1 mm)
Specific heat (J/kg/K)	Coil 380, Stator core 492, Magnetic ring 560 Shaft 467, Rotor core 467, Permanent magnet 430,
Heat transfer coefficient (W/ K·m^2)	Surface in contact with air 10, Surface in contact with the air gap 340.3

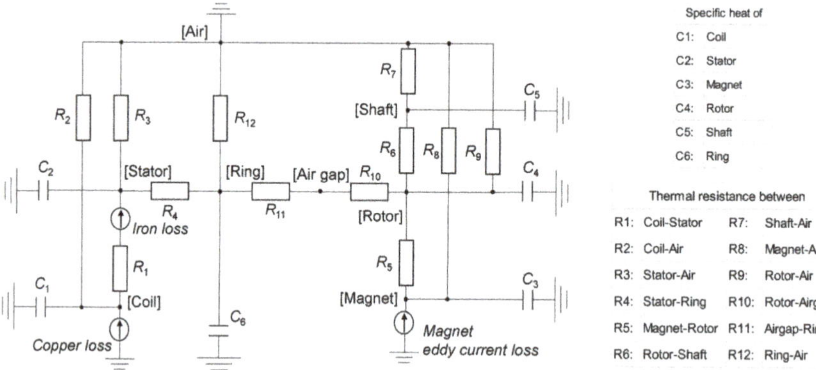

Figure 4. Simulation motor thermal circuit.

3.2. Loss Analysis Results

Figure 5 shows the magnetic flux density distribution calculated by simulation. In open-slot motors, most of the magnetic flux passes through the teeth of the stator, but in closed-slot motors, the short circuit of the magnetic flux for the ring is quite large. Conversely, the short circuit of the ring is small for the ring motor. Figure 6 shows the angle dependence of the magnetic flux density in the gap. The open-slot motor is close to a square wave, whereas the closed-slot motor is almost sinusoidal. The ring motor is also in a state approximal to a sine wave. Figure 7 shows the Fourier transform result for the magnetic flux density, from which it is evident that harmonic components are significantly suppressed and the fundamental wave component is increased in closed-slot motors, which is consistent with the tendency in Figure 6a. Moreover, for the ring motors, the fourth, fifth and seventh harmonic components decrease by 50% or more. Accordingly, suppression of spatial harmonics can be realized using a ring.

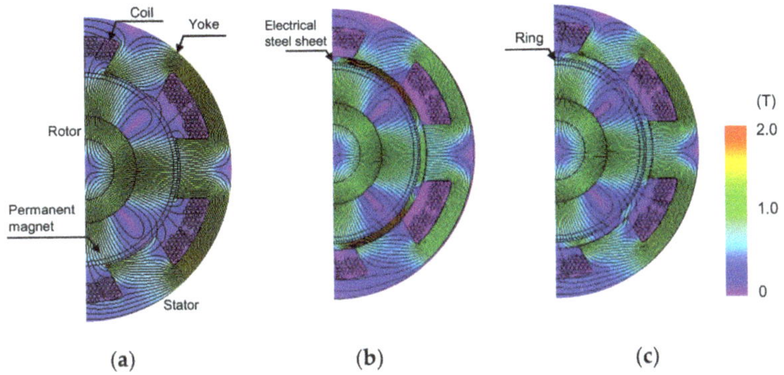

Figure 5. Magnetic flux density distributions (N = 30,000 rpm, P_o = 500 W): (**a**) open-slot motor, (**b**) closed-slot motor, and (**c**) ring motor.

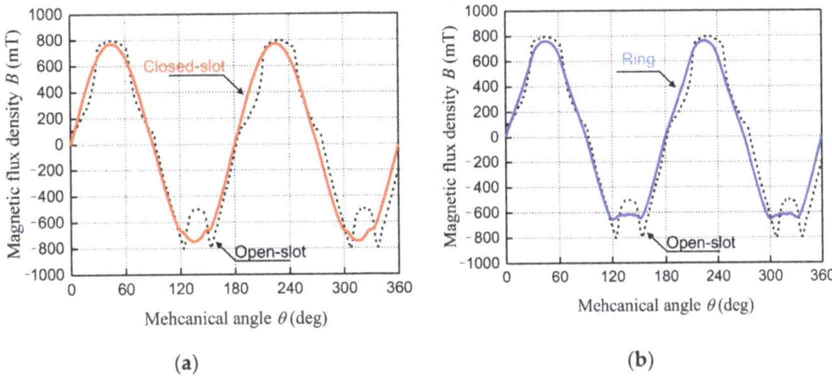

Figure 6. Magnetic flux density at gap (N = 30,000 rpm, P_o = 500 W): (**a**) comparison of open-slot and closed-slot motor and (**b**) comparison of open-slot and ring motor.

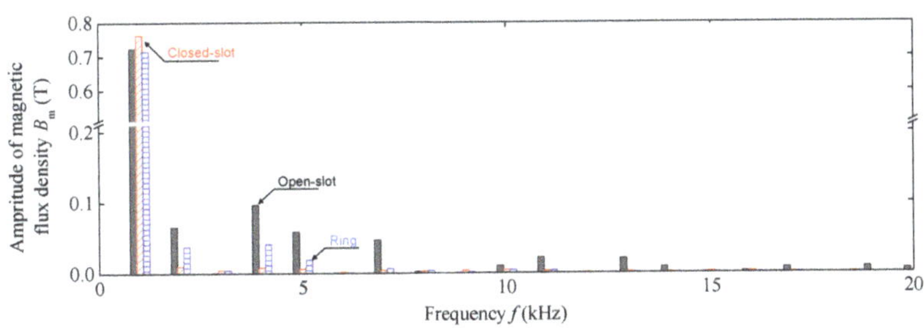

Figure 7. Spectrum of magnetic flux density (N = 30,000 rpm, P_o = 500 W): (**a**) open-slot motor, (**b**) closed-slot motor, and (**c**) ring motor.

Figure 8 shows the current density distribution. The magnet of an open-slot motor has a large area of high current density. However, the area of high current density is small in the magnet part of magnetic ring motors and closed-slot motors. The generation of eddy currents in the magnet can be reduced with a ring motor because the spatial harmonics are suppressed, as mentioned in the previous section.

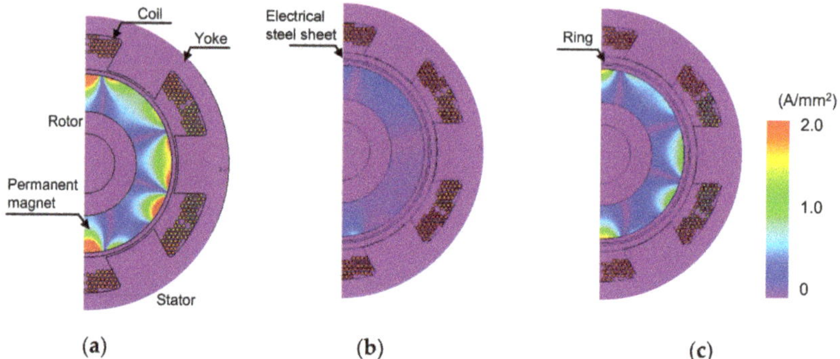

Figure 8. Current density distributions ($N = 30{,}000$ rpm, $P_o = 500$ W): (**a**) open-slot motor, (**b**) closed-slot motor, and (**c**) ring motor.

Figure 9 shows the torque characteristics of each motor. Indeed, the decrease in the torque constant of the closed-slot motor is remarkable. In particular, the closed-slot motors have a torque constant that is 18% smaller than that of open-slot motors. Although the torque characteristics are reduced by providing the ring, the torque constant is reduced by 8% in the ring motor compared with the closed-slot motor. The short circuit of the magnetic flux of the magnet was suppressed, and the reduction in the effective magnetic flux that contributes to the torque was avoided by applying a composite material with low relative permeability to the ring as shown in Figure 3.

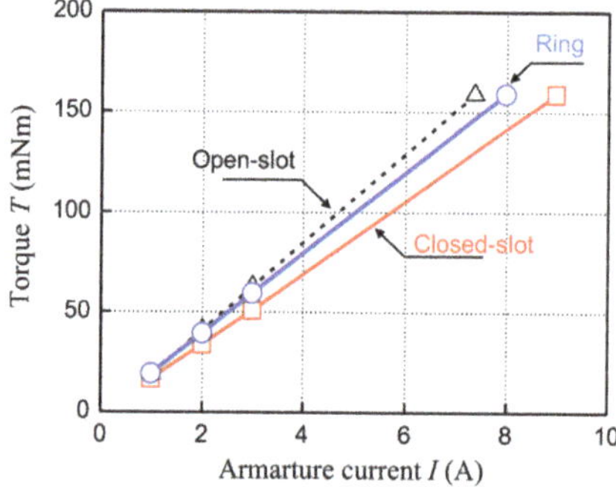

Figure 9. Thrust characteristics.

Table 5 shows the results of comparing the losses of each motor. The open-slot motor has a large magnet eddy current loss. The closed-slot motor can reduce the eddy current loss of the magnet, but the copper loss increases significantly as the torque constant decreases for an output power of 500 W. Although the torque constant of the magnetic ring motor is slightly reduced, the expected effect of the magnet eddy current loss is reduced by 78%. The ring motor had the highest efficiency compared with other motors operating at the same output of 500 W. A bond magnet motor with the same dimensions as the open-slot motor was also simulated for comparison to continue torque characteristics and losses. A bond magnet (Hidence-1000) was used as the permanent magnet of the open-slot motor. The magnet eddy current loss of the bond magnet motor was reduced. However, the torque constant is considerably lower than that of the magnetic ring motor, and the total loss is also large. Accordingly, the magnetic ring can reduce the magnet eddy current loss without significantly reducing the torque constant.

Table 5. Motor loss analysis results (N = 30,000 rpm, P_o = 500 W).

Item	(a) Open-Slot Motor	(b) Closed-Slot Motor	(c) Ring Motor	(d) Bonded Magnet Motor
Torque constant (mNm/A)	30.5	25.1	28.2	8.9
Copper loss (W)	19.8	29.2	23.2	51.6
Iron loss in the stator (W)	13.4	15.0	10.0	1.9
Eddy current loss in the permanent magnet (W)	8.5	1.0	1.9	0.1
Iron loss in the ring (W)	-	-	0.8	-
Total loss (W)	41.7	45.2	35.9	53.6
Efficiency (%)	91.7	91.0	92.8	89.3

Figure 10 shows the dependence of the efficiency on the output in each motor. Magnetic ring motors maintain higher efficiency than open-slot motors from low- to high-power regions. Since the closed-slot motor have a low torque constant, copper loss increases at high output. Conversely, since the torque constant of the ring motor is higher than that of the closed-slot motor, the efficiency is the highest due to the effect of suppressing copper loss and reducing the eddy current loss of the magnet at high output. The ring motor achieves high efficiency in the high speed and torque range.

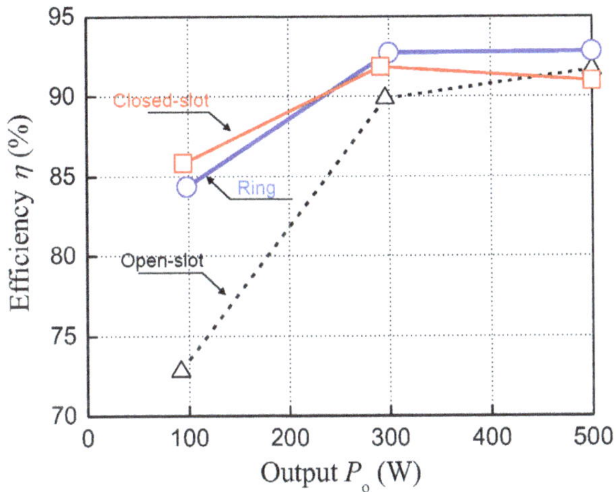

Figure 10. Dependence of the motor efficiency on the output.

3.3. Temperature Rise Analysis Results

Mounting a ring is effective in reducing the eddy current loss of the magnet. Therefore, the effect of reducing the rotor temperature due to the magnetic ring has been confirmed by thermal analysis simulation. Figure 11 shows the time transition of the temperature rise of each part. Overall, the saturation temperature rise of magnets for closed-slot motors and ring motors is reduced by over 30 K. Particularly, the saturation temperature rise at 30,000 rpm with 500 W is lower than 40 K for the open-slot and the ring motor. The temperature rise of the magnet of the closed-slot motor reduced to the same value as that of the ring motor. Accordingly, a magnetic composite ring is effective in reducing the temperature rise of permanent magnets. Ring motors can be also expected to suppress thermal demagnetization and magnet breakage.

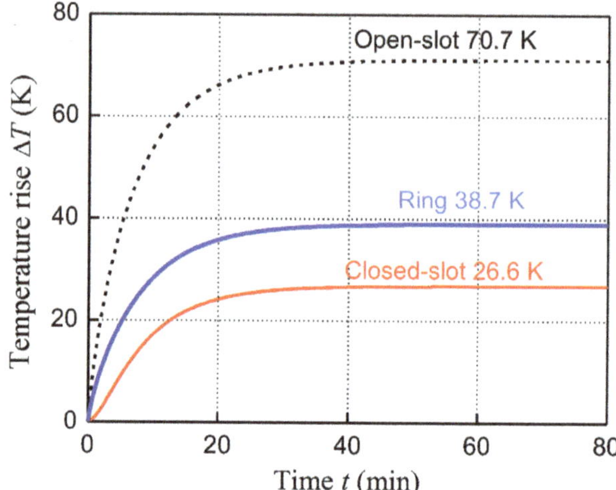

Figure 11. Time transition of temperature rise at magnets (N = 30,000 rpm, P_o = 500 W).

4. Torque Characteristics

In Section 3, the magnetic ring was shown to be effective in suppressing spatial harmonics. Alternatively, for the closed-slot motor, the magnetic flux of the magnet often short-circuited, and the torque constant was significantly reduced. A short circuit of the magnetic flux occurs on the stator side, and the torque constant decreases due to the characteristics of the magnetic ring. Therefore, this section focuses on the thickness and relative permeability of the magnetic ring, which are influencing factors over the short circuiting of the magnetic flux. Simulation results and magnetic circuit theory are used to explain the effect of changing the ring thickness and relative permeability on the torque characteristics.

4.1. Torque Characteristics

The relative permeability and thickness of the ring affect the short-circuit state of the permanent magnet magnetic flux on the stator side, by which the torque constant changes. Therefore, the torque characteristics when the thickness of the magnetic ring and the relative permeability are changed were derived by magnetic field simulation.

The thickness of the magnetic ring was changed to 0.5, 0.6, 0.7, 0.9, and 1.1 mm to confirm the effect of short circuiting the permanent magnet flux on the stator side. The outer diameter of each motor was 32 mm, the air gap was 0.7 mm, and the thickness of the magnet changed according to changes in the magnetic ring thickness. The relative permeability of the ring was constant at 10. All motor outputs were 500 W, and the simulation conditions are the same as those in Section 3. Figure 12a,b show the dependency on ring thickness of the torque constant, eddy current loss of the magnet, and motor efficiency based on the simulation. As the thickness of the magnetic ring increases, the torque constant decreases proportionally. Conversely, the eddy current loss of the magnet increases as the ring thickness increases. Maximum efficiency is achieved with a ring thickness of 0.7 mm, which can be attributed to the trade-off between copper loss and eddy current loss in the magnets. The copper loss of the coil increases because the current required for the same output increases as the torque constant decreases.

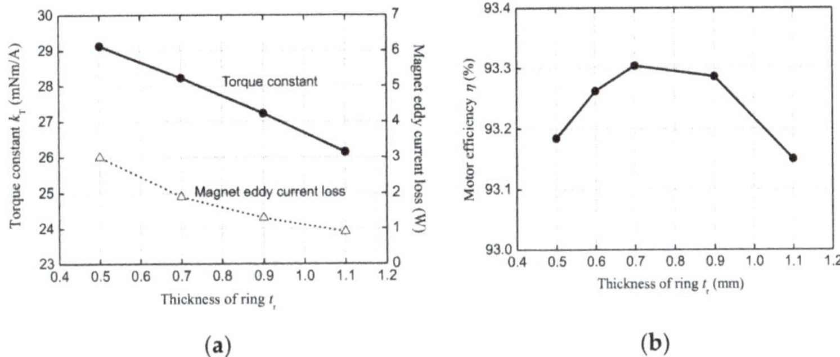

(a)   (b)

Figure 12. Effect of changing ring thickness (N = 30,000 rpm, P_o = 500 W): (**a**) torque constant and magnet eddy current loss and (**b**) motor efficiency.

The torque characteristics were compared by changing the relative magnetic permeability to 10, 60, and 1000 with a ring thickness of 0.7 mm. Figure 13a,b show the dependency on ring relative permeability of the torque constant, eddy current loss of the magnet, and motor efficiency based on the simulation. Rings with a relative permeability of 10 are used in the transformer cores of DC-DC converters [38]. A permeability of 60 and 1000 was assumed for the dust core manufactured by Dongbu Electronic Materials Co., Ltd. and for the electrical steel sheet, respectively. The torque constant decreases in inverse proportion; conversely, the magnet eddy current loss reduced with an increase in the relative magnetic permeability. The efficiencies for the relative permeabilities of 10 and 60 are almost the same, but the effect of lowering the torque constant is greater than that of reducing the eddy current loss at a relative permeability of 1000. Accordingly, to obtain the same output of 500 W, the copper loss increases to a large value and the efficiency decreases.

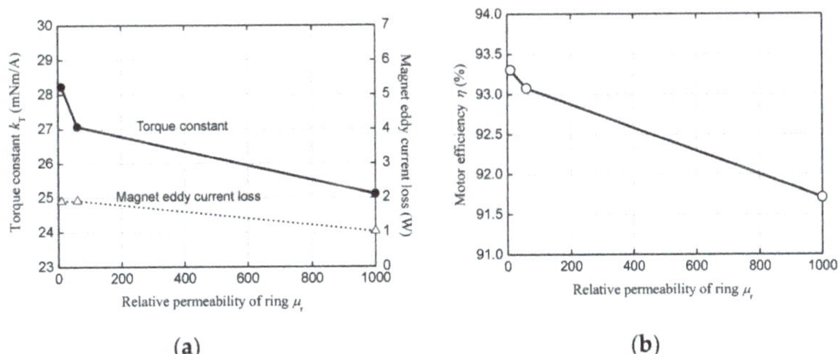

(a)   (b)

Figure 13. Effect of changing ring relative permeability (N = 30,000 rpm, P_o = 500 W): (**a**) torque constant and magnet eddy current loss and (**b**) motor efficiency.

4.2. Magnteic Circuit

The simulation results in Section 4.1 are analyzed by comparing them with the magnetic circuit of the motor. Figure 14 shows a magnetic circuit of a magnetic ring motor. R_m, R_a, R_r, and R_s denote the reluctance of the magnet, the air gap, the ring, and the stator core, respectively. Although the cross-sectional area through which magnetic flux passes differs between the teeth and yoke of the stator core, the magnetic resistance is smaller than that of the ring. Therefore, the reluctance of the teeth and yoke of the stator core can be calculated together for simplification. The magnetic resistance of each part is given by Equation (1).

$$R = \frac{l}{\mu\mu_0 dw} \tag{1}$$

where μ denotes the relative permeability, l denotes the magnetic path length, w denotes the axial length, and d denotes the magnetic path cross-sectional length.

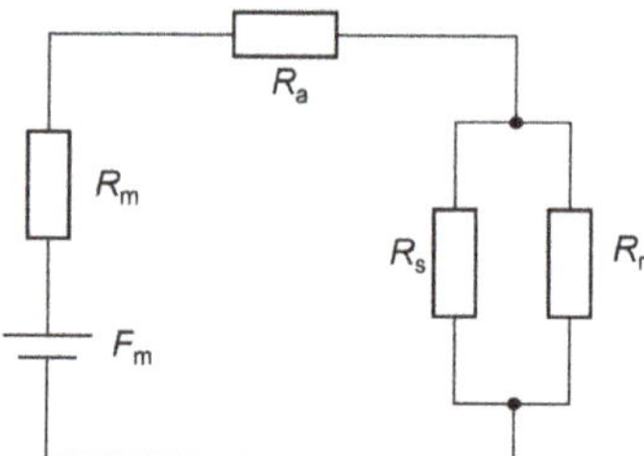

Figure 14. Ring motor magnetic circuit.

The relationship between the short-circuit magnetic flux of the ring (ϕ_r), the stator magnetic flux that contributes to torque (ϕ_s), and the magnet magnetic flux (ϕ_m) is given by Equation (2). The magnetomotive force (F_m) of a magnet is represented by the product of the coercive force (H_m) of the magnet and the thickness of the magnet (t_m) as described in Equation (3). The magnet thickness t_m changes in proportion to ring thickness (t_r), as described by Equation (4), assuming that the motor external dimensions and air gap distance are constant in this study. Therefore, magnet magnetic flux changes in proportion to the thickness of the magnetic ring, as described by Equation (5) with total reluctance (R_t). Equation (6) expresses the relationship between the magnet magnetic flux and the stator magnetic flux in the shunt rule of the magnetic circuit with reluctances R_s and R_r.

The stator magnetic flux is represented by the product of the term related to the shunt flow of the ring and the term related to the magnet magnetic flux. The term of the magnet magnetic flux directly decreases more than the term related to the ring shunt flow with same the relative magnetic permeability of the ring. Therefore, the torque magnetic flux decreases in proportion to magnet thickness.

When the magnet thickness is increased with the same relative magnetic permeability of the ring, the influence of the magnetomotive force term is larger than that related to the shunt flow. Therefore, the stator magnetic flux ϕ_s decreases, and the torque constant proportionally decreases, which is the same relationship shown in Figure 12a. Conversely, when the relative permeability is increased while maintaining the same magnetic ring thickness, only the shunt term of the ring decreases in an inverse proportion. Therefore, the stator magnetic flux decreases, and the torque constant decreases in an inverse proportion, which is the same relationship shown in Figure 13a.

$$\phi_m = \phi_r + \phi_s \tag{2}$$

$$F_m = H_m t_m \tag{3}$$

$$t_m = \frac{D_o}{2} - \frac{D_f}{2} - t_g - t_s - t_r \tag{4}$$

$$\phi_m = \frac{F_m}{R_t} \propto -t_r \tag{5}$$

$$\phi_s = \frac{R_r}{R_s + R_r} \times \phi_m = \frac{\frac{l_r}{\mu_r \mu_0 t_r w_m}}{\frac{l_s}{\mu_s \mu_0 d_s w_m} + \frac{l_r}{\mu_r \mu_0 t_r w_m}} \times \phi_m = \frac{\mu_s d_s l_r}{\mu_r t_r l_s + \mu_s d_s l_r} \times \phi_m \tag{6}$$

where ϕ_r: the short-circuit magnetic flux of the ring, ϕ_s: the stator magnetic flux that contributes to torque, ϕ_m: the magnet magnetic flux, F_m: magnetomotive force of a magnet, H_m: coercive force, t_m: thickness of the magnet, D_o: stator outer diameter, D_f: shaft outer diameter, t_g: air gap length, t_s: thickness of electrical steel sheet, t_r: thickness of ring, R_t: synthetic reluctance, l_s: magnetic flux length through the stator, d_s: width of stator teeth, w_m: axial length of the motor, l_r: magnetic flux length through the ring, μ_r: relative permeability of ring, and μ_s: relative permeability of stator.

The magnetic resistance of the ring decreases by larger the relative permeability or thickness of the ring, which increases the short-circuiting of the magnetic flux. This means that the stator magnetic flux decreases in proportion to the torque constant k_t. Therefore, setting an appropriate magnetic ring thickness and relative magnetic permeability based on the motor specifications, such as the magnetomotive force and stator size, is important with respect to improving motor efficiency by using the magnetic ring.

5. Conclusions

A concentrated winding motor that tends to generate spatial harmonics has a high magnet eddy current loss at high-speed rotation. This paper proposes a method of attaching a resin ring mixed with magnetic powder to the stator to suppress spatial harmonics. We have clarified the following points by simulation.

(1) The fourth, fifth, and seventh harmonic components were reduced by over 50% using a 0.7 mm-thick composite ring.
(2) The magnet eddy current loss for the ring motor was reduced by 78%, and the temperature rise value was reduced by 30 K compared with the open-slot motor.
(3) The efficiency of the ring motor was improved by 1.1% and 1.8% compared with the open-slot and closed-slot motor made of electrical steel sheeting, respectively.

The closed-slot motor was the most effective in suppressing spatial harmonics, but the torque constant was significantly reduced due to the high relative magnetic permeability of the electromagnetic steel sheets forming the ring. Therefore, the effects of ring thickness and relative magnetic permeability on torque characteristics were confirmed by simulations and magnetic circuits.

Increasing the magnetic flux of the magnet passing through the ring reduced the torque constant of the motor. Setting both the appropriate composite ring thickness and relative permeability is essential with respect to suppressing spatial harmonics without compromising torque characteristics.

Author Contributions: Conceptualization and methodology, M.S., M.N., T.M. and Y.B.; software, K.T., M.H., R.Y. and R.M.; validation, K.T., R.Y. and R.M.; writing—original draft preparation, M.S. and K.T.; funding acquisition, M.S. and T.M. All authors have read and agreed to the published version of the manuscript.

Funding: This research is supported by Adaptable and Seamless Technology transfer Program through Target-driven R&D (A-STEP) from Japan Science and Technology Agency (JST).

Conflicts of Interest: The authors declare no conflict of interest.

References

1. Lee, J.; Jung, D.; Lim, J.; Lee, K.; Lee, J. A study on the synchronous reluctance motor design for high torque by using RSM. *IEEE Trans. Magn.* **2018**, *54*, 1–5. [CrossRef]
2. Liu, C.; Shih, P.; Cai, Z.; Yen, S.; Lin, H.; Hsu, Y.; Luo, T.; Lin, S. rotor conductor arrangement designs of high-efficiency direct-on-line synchronous reluctance motors for metal industry applications. *IEEE Trans. Ind. Appl.* **2020**, *56*, 4337–4344.
3. Tabora, J.M.; de Lima Tostes, M.E.; de Matos, E.O.; Soares, T.M.; Bezerra, U.H. Voltage harmonic impacts on electric motors: A comparison between IE2, IE3 and IE4 induction motor classes. *Energies* **2020**, *13*, 3333. [CrossRef]
4. Gómez, J.R.; Quispe, E.C.; Castrillón, R.P.; Viego, P.R. Identification of technoeconomic opportunities with the use of premium efficiency motors as alternative for developing countries. *Energies* **2020**, *13*, 5411. [CrossRef]
5. Son, J.-C.; Kang, Y.-R.; Lim, D.-K. Optimal design of IPMSM for FCEV using novel immune algorithm combined with steepest descent method. *Energies* **2020**, *13*, 3395. [CrossRef]
6. Cui, S.; Zhao, T.; Du, B.; Cheng, Y. Multiphase PMSM with asymmetric windings for electric Drive. *Energies* **2020**, *13*, 3765. [CrossRef]

7. de Pinto, S.; Camocardi, P.; Chatzikomis, C.; Sorniotti, A.; Bottiglione, F.; Mantriota, G.; Perlo, P. On the comparison of 2- and 4-wheel-drive electric vehicle layouts with central motors and single- and 2-speed transmission systems. *Energies* **2020**, *13*, 3328. [CrossRef]

8. Flieh, H.M.; Slininger, T.; Lorenz, R.D.; Totoki, E. Self-sensing via flux injection with rapid servo dynamics including a smooth transition to back-EMF tracking self-sensing. *IEEE Trans. Ind. Appl.* **2020**, *56*, 2673–2684. [CrossRef]

9. Mese, E.; Yasa, Y.; Ertugrul, B.; Sincar, E. Design of a high performance servo motor for low speed high torque application. In Proceedings of the 2014 International Conference on Electrical Machines (ICEM), Berlin, Germany, 2–5 September 2014; pp. 2014–2020. [CrossRef]

10. Ogata, T.; Maruyama, H.; Fujita, K. A spindle motor servo using a disturbance observer. In Proceedings of the 2009 Digest of Technical Papers International Conference on Consumer Electronics, Las Vegas, NV, USA, 10–14 January 2009; pp. 1–2. [CrossRef]

11. Liu, N.-W.; Hung, K.-Y.; Yang, S.-C.; Lee, F.-C.; Liu, C.-J. Design of high-speed permanent magnet motor considering rotor radial force and motor losses. *Energies* **2020**, *13*, 5872. [CrossRef]

12. Kim, S.; Kim, Y.; Lee, G.; Hong, J. A novel rotor configuration and experimental verification of interior PM synchronous motor for high-speed applications. *IEEE Trans. Magn.* **2012**, *48*, 843–846. [CrossRef]

13. Wang, X.; Luo, J.; Du, M. Rotor strength analysis of high-speed motor based on magnetic pole eccentricity. *AIP Conf. Proc.* **2019**, *2154*, 020008. [CrossRef]

14. Frosini, L.; Pastura, M. Analysis and design of innovative magnetic wedges for high efficiency permanent magnet synchronous machines. *Energies* **2020**, *13*, 255. [CrossRef]

15. Atallah, K.; Howe, D.; Mellor, P.H.; Stone, D.A. Rotor loss in permanent magnet brushless AC machine. *IEEE Trans. Ind. Appl.* **2000**, *36*, 1612–1618.

16. Jung, J.-W.; Lee, B.-H.; Kim, K.-S.; Kim, S.-I. Interior permanent magnet synchronous motor design for eddy current loss reduction in permanent magnets to prevent irreversible demagnetization. *Energies* **2020**, *13*, 5082. [CrossRef]

17. Zhu, Z.Q.; Howe, D. Influence of design parameters on cogging torque in permanent magnet machines. *IEEE Trans. Energy Convers.* **2000**, *15*, 407–412. [CrossRef]

18. Kaimori, H.; Akatsu, K. Behavior Modeling of permanent magnet synchronous motors using flux linkages for coupling with circuit simulation. *IEEJ J. Ind. Appl.* **2018**, *7*, 56–63. [CrossRef]

19. Li, Q.; Zhang, B.; Liu, A. Cogging torque computation and optimization in dual-stator axial flux permanent magnet machines. *IEEJ Trans. Electr. Electron. Eng.* **2020**, *15*, 1414–1422. [CrossRef]

20. Jo, I.; Lee, H.; Jeong, G.; Ji, W.; Park, C. A study on the reduction of cogging torque for the skew of a magnetic geared synchronous motor. *IEEE Trans. Magn.* **2019**, *55*, 1–5. [CrossRef]

21. Mitsui, Y.; Ahmed, S.; Aoyama, Y.; Koseki, T. Detent force reduction by positional shifting of permanent magnets for a PMLSM without compromising thrust. *IEEJ J. Ind. Appl.* **2020**, *9*, 376–383.

22. Tang, H.; Zhang, M.; Dong, Y.; Li, W.; Li, L. Influence of the opening width of stator semi-closed slot and the dimension of the closed slot on the magnetic field distribution and temperature field of the permanent magnet synchronous motor. *IET Electr. Power Appl.* **2020**, *14*, 1642–1652. [CrossRef]

23. Chun, J.-S.; Jung, H.-K.; Yoon, J.-S. Shape optimization of closed slot type permanent magnet motors for cogging torque reduction using evolution strategy. *IEEE Trans. Magn.* **1997**, *33*, 1912–1915. [CrossRef]

24. Tsunata, R.; Takemoto, M.; Ogasawara, S.; Watanabe, A.; Ueno, T.; Yamada, K. Development and evaluation of an axial gap motor using neodymium bonded magnet. *IEEE Trans. Ind. Appl.* **2018**, *54*, 254–262. [CrossRef]

25. Hosoya, R.; Shimomura, S. Apply to in-wheel machine of permanent magnet vernier machine using NdFeB bonded magnet—Fundamental study. In Proceedings of the 8th International Conference on Power Electronics—ECCE Asia, Jeju, Korea, 30 May–3 June 2011; pp. 2208–2215. [CrossRef]

26. Shi, W.; Zhou, X. Online estimation method for permanent magnet temperature of high-density permanent magnet synchronous motor. *IEEJ Trans. Electr. Electron. Eng.* **2020**, *15*, 751–756. [CrossRef]

27. Sato, M.; Sugimoto, K.; Kubota, K.; Endo, S.; Mizuno, T. Reducing the alternating current resistance and heat generation in a single-wire coil using a magnetic tape. *IEEJ Trans. Electr. Electron. Eng.* **2020**, *15*, 1541–1548. [CrossRef]

28. Guo, Y.; Zhu, J.; Lu, H.; Lin, Z.; Li, Y. Core loss calculation for soft magnetic composite electrical machines. *IEEE Trans. Magn.* **2012**, *48*, 3112–3115. [CrossRef]

29. Guo, Y.; Zhu, J.G.; Watterson, P.A.; Wu, W. Development of a PM transverse flux motor with soft magnetic composite core. *IEEE Trans. Energy Convers.* **2006**, *21*, 426–434. [CrossRef]

30. Guo, Y.; Zhu, J.; Dorrell, D.G. Design and analysis of a claw pole permanent magnet motor with molded soft magnetic composite core. *IEEE Trans. Magn.* **2009**, *45*, 4582–4585.

31. Enomoto, Y.; Ito, M.; Koharagi, H.; Masaki, R.; Ohiwa, S.; Ishihara, C.; Mita, M. Evaluation of experimental permanent-magnet brushless motor utilizing new magnetic material for stator core teeth. *IEEE Trans. Magn.* **2005**, *41*, 4304–4308. [CrossRef]

32. Shutta, Y.; Ako, Y.; Takahashi, Y.; Fujiwara, K.; Kitao, J.; Yamada, M. Conductivity and AC magnetic loss of Nd–Fe–B sintered magnet. *Int. J. Appl. Electromagn. Mech.* **2019**, *61*, S3–S12. [CrossRef]

33. Li, Y.; Jiang, B.; Zhang, C.; Liu, Y.; Gong, X. Eddy-current loss measurement of permanent magnetic material at different frequency. *Int. J. Appl. Electromagn. Mech.* **2019**, *61*, S13–S22. [CrossRef]

34. Yamazaki, K.; Isoda, Y. Iron loss and magnet eddy current loss analysis of IPM motors with concentrated windings. *IEEJ Trans. Ind. Appl.* **2008**, *128*, 678–684. [CrossRef]

35. Suzuki, T.; Nirei, M.; Yamamoto, T.; Yamanaka, Y.; Goto, T.; Sato, M.; Bu, Y.; Mizuno, T. Reduction of eddy currents in winding by using a magnetic layer on an IPM motor. *IEEJ Trans. Electr. Electron. Eng.* **2020**, *15*, 601–606. [CrossRef]

36. Sato, M.; Nirei, M.; Yanamaka, Y.; Suzuki, T.; Bu, Y.; Mizuno, T. Increasing the efficiency of a drone motor by arranging magnetic sheets to windings. *Energy Rep.* **2020**, *6*, 439–446. [CrossRef]

37. JMAG. Available online: https://www.jmag-international.com/ (accessed on 14 December 2020).

38. Yabu, N.; Sugimura, K.; Sonehara, M.; Sato, T. Fabrication and evaluation of composite magnetic core using iron-based amorphous alloy powder with different particle size distributions. *IEEE Trans. Magn.* **2018**, *54*, 1–5. [CrossRef]

Publisher's Note: MDPI stays neutral with regard to jurisdictional claims in published maps and institutional affiliations.

MDPI

St. Alban-Anlage 66

4052 Basel

Switzerland

Tel. +41 61 683 77 34

Fax +41 61 302 89 18

www.mdpi.com

Energies Editorial Office

E-mail: energies@mdpi.com

www.mdpi.com/journal/energies

www.ingramcontent.com/pod-product-compliance
Lightning Source LLC
LaVergne TN
LVHW070651170726
843515LV00004B/380